The Emergence of Bicycling and Automobility in Britain

The Emergence of Bicycling and Automobility in Britain

Craig Horner

BLOOMSBURY ACADEMIC
LONDON • NEW YORK • OXFORD • NEW DELHI • SYDNEY

BLOOMSBURY ACADEMIC
Bloomsbury Publishing Plc
50 Bedford Square, London, WC1B 3DP, UK
1385 Broadway, New York, NY 10018, USA
29 Earlsfort Terrace, Dublin 2, Ireland

BLOOMSBURY, BLOOMSBURY ACADEMIC and the Diana logo are trademarks
of Bloomsbury Publishing Plc

First published in Great Britain 2021
This paperback edition published in 2022

Copyright © Craig Horner, 2021

Craig Horner has asserted his right under the Copyright, Designs and Patents Act, 1988,
to be identified as Author of this work.

For legal purposes the Acknowledgements on p. ix constitute an extension
of this copyright page.

Cover design: Terry Woodley
Cover image: C. Jarrott, C. G. Wridgway and S. F. Edge at the start of the Motor-Car Club
Championship in 1899. Universal Images Group North America LLC / Alamy Stock Photo

All rights reserved. No part of this publication may be reproduced or transmitted in
any form or by any means, electronic or mechanical, including photocopying,
recording, or any information storage or retrieval system, without prior
permission in writing from the publishers.

Bloomsbury Publishing Plc does not have any control over, or responsibility for, any
third-party websites referred to or in this book. All internet addresses given in this
book were correct at the time of going to press. The author and publisher regret any
inconvenience caused if addresses have changed or sites have ceased to exist,
but can accept no responsibility for any such changes.

A catalogue record for this book is available from the British Library.

A catalog record for this book is available from the Library of Congress.

ISBN: HB: 978-1-3500-5418-9
        PB: 978-1-3502-1456-9
        ePDF: 978-1-3500-5420-2
        ePub: 978-1-3500-5421-9

Typeset by Deanta Global Publishing Services, Chennai, India

To find out more about our authors and books visit www.bloomsbury.com
and sign up for our newsletters.

This book is dedicated to the memory of Malcolm Jeal (1944–2016)

Contents

List of figures		viii
Acknowledgements		ix
1	Introduction	1
2	Resistance to change	13
3	Entrepreneurs	41
4	Trials	65
5	The 'old brigade' and the new 'steady and careful artisan'	89
6	Tourists	121
7	Futures	143
8	Conclusion	169
Bibliography		181
Appendix: Biography		195
Index		198

Figures

2.1	A cycle club out on tour, *c.* 1880, with a mixture of 'ordinaries' and tricycles	32
2.2	The eliminating trials for the 1903 Gordon Bennett Cup held in the grounds of Welbeck Abbey, Notts	36
3.1	*Punch* magazine initially lampooned the motor car as uncontrollable, and the motorist as hapless	43
3.2	S. F. Edge, repairs a puncture at the roadside on the Napier during the Thousand Mile Trial	51
3.3	S. F. Edge on a 1901 De Dion tricycle with an air-cooled 2¾hp engine	52
4.1	The Thousand Mile Trial of 1900 stopped off in Manchester, permitting respectable-looking members of the public to inspect the vehicles	67
4.2	Vehicles on display in the Royal Botanical Gardens at Old Trafford in Manchester during the Thousand Mile Trial	70
4.3	Charles Jarrott and S. F. Edge pose on the 6hp Panhard-Levassor (known as Number 8)	83
5.1	S. F. Edge endorsing an Ariel bicycle	92
5.2	Harry Lawson's 'Gyroscope' of *c.* 1902, demonstrated by his daughters	94
5.3	Charles Jarrott in the driving seat of a Leon Bollée	96
5.4	A motor-cycling couple, oblivious to the foul weather	99
5.5	Major C. G. Matson in the driving seat of his De Dion, about 1903	102
5.6	The 'Flying Pennington Cycle' (1896) is made to leap a ravine of sixty-five feet	111
6.1	Motor cars at Ascot, 1900, the first year they were allowed into the enclosure	124
7.1	*Punch* magazine quickly changed the way it depicted motoring, moving to portray the motorist and motor car as encroaching and menacing	150
7.2	'Almost a motorcar!' The 'trimo' promised motor-car luxury at the price of a motor-cycle	156
7.3	The cyclecar ('new motoring') was a phenomenon of the period immediately before the First World War, and provided very cheap and cheerful motoring	158
8.1	'An enjoyable experience: Their first trial trip'. Many such drawings promised the open road and more for the courting couple	175
8.2	The annual dinner of the Fellowship of Old-Time Cyclists, 1927	178

Acknowledgements

Twelve or so years on, this book has now appeared in large part because of the many good people whose better nature I have taken advantage of. I'd like to thank them all here, but there are so many that I fear I will overlook a few, and to those I apologize now.

But for those I can identify, I offer my heartfelt thanks. At Manchester Metropolitan University, Professors Alan Kidd and Melanie Tebbutt provided moral and friendly support, particularly as I risked professional oblivion by changing my area of expertise (formerly, eighteenth-century British society). Also, over the years, although they probably didn't realise it at the time, I have drawn support and confidence through conversations with Gijs Mom, Massimo Moraglio, Peter Norton, Rich Stephens and Nick Clayton.

In the process, several people have passed away who had been very helpful. I just wasn't fast enough for them to be able to read the final version. Above all this includes Malcolm Jeal, but also Nick Georgano, John Warburton, Sandy Skinner, Peter Heilbron and Derek Grossmark.

I called on many librarians, archivists and editors, in particular Richard Roberts, Ian Ferguson, Ian Sykes, Jane Holmes, Danny Smith, Nic Ward, Nigel Land and the estimable Patrick Collins.

Two anonymous reviewers, plus Anders Clausager, who read an early draft, have also saved me from delivering a bit of a calamity. Here, too, fellow motoring historians Mike Worthington Williams, David Burgess-Wise, Tom Clarke, Peter Moss, Peter Card, Simon Fisher, Thomas Ulrich, Michael Edwards, Malcolm Bobbitt, John Harrison and Josh Butt have all been generous with their expertise.

I also thank Des Donohoe, the Revd Canon Jeffrey Bell, Rosemary Sharples, Helen O'Neill, Peter Jackson, Jonathan Rishton, Damien Kimberley, Corey Estensen, Eunice Jeal and Julia Dawson. Correspondence and meetings with some descendants of S.F. Edge and Charles Jarrott – Michael and June Cooke, Penny Morris and Sue Edge – have been immensely rewarding.

Of particular help, providing me the run of their superb resources, were the Vintage Sports-Car Club, the Veteran-Cycle Club, the Reference Library at the National Motor Museum, Beaulieu, the collections of the Royal Automobile Club (at Westminster Archive Centre), and the Richard Roberts Archive in Stockport. I also took advantage of access to the Museum of English Rural Life in Reading, the Modern Records Centre at Warwick University, the BFI Reuben library, Penge Library and the British Library. The excellent gracesguide.co.uk was used to help fill in the gaps for the biographies.

My apologies to the many whose hopes I raised over the years that this would be a biography of S. F. Edge. That, as you'll see, remains to be written. Meanwhile, all errors here are mine.

And above all, thanks and love to Nancy, Martha and Judith.

1

Introduction

The appearance of the motor car – or, in the then new legal terminology, 'light locomotive' – on the public highways of the United Kingdom in the 1890s gave advertisers a new medium to promote their products. This happened to be particularly the case with soap. Mr Goodwin advertised Mother Shipman's Soap as he drove his Lutzmann Benz around Manchester in about 1896[1] – his was probably the first motor car in the town, so he could be sure of attention. Similarly, Lever Brothers used the novelty of 'motor vans' around the country to advertise their Sunlight soap in 1896; this must have presented a sight sufficiently common for the *Illustrated London News* to feel able to publish a cartoon that same year featuring Mr and Mrs John Bull as they surveyed a passing 'Sunlight' van.[2]

In the popular imagination, then, the motor car could be little more than a freakish contraption with few uses beyond advertising. This presented a challenge for entrepreneurs trying to promote the new 'motor traction' as a potential new industry and practical means of transport in its own right. The journalist Henry Sturmey (1857–1930), at that time the editor of *The Autocar* magazine, knew this when he set out in 1897 to be the first to drive a motor car from John O'Groat's to Land's End. His suspicion would have been reinforced by a conversation he overheard. He was at John O'Groat's House with his Daimler motor car when a 'coach and four [horses]' delivered a 'fashionably-dressed party' of gentlemen. Having spotted the Daimler, one said, 'I wonder what that motor-car was doing at Groat's today?' 'Oh, another beastly advertisement – Pear's soap, or something of that kind, I suppose.' Sturmey knew what this meant: no gentleman would 'demean himself by being seen on one [a motor car] so long as there was a good horse about'.[3]

Before the turn of the twentieth century, and for some years beyond, motoring played no useful part in the lives of most of the population. For travelling beyond their locality, middle-class families might have kept a horse, or hired one as necessary, possibly with a carriage. The 'safety' bicycle was becoming popular, while the train was affordable, sensible and fast. The steam traction engine had been present on the public

---

[1] Science and Industry Museum Archive, YMS 0197.2.
[2] *Illustrated London News*, 28 November 1896, p. 729.
[3] Henry Sturmey, 'The amazing growth of the motoring movement: a contrast between fifteen years ago and now', *The Motor*, 11 March 1913, pp. 239–40. See his account of the trip written at the time: Henry Sturmey, *On an Autocar through the Length and Breadth of the Land* (London: Illife, Sons and Sturmey, 1898).

highway for half a century or more,[4] but many people had yet to see a light motor car, and this remained the case into the first decade of the twentieth century. A cartoon in *Punch* magazine in 1902 illustrates this. With a milestone indicating they were well into the Wiltshire countryside, a girl in a horse and carriage, seeing a motor car for the first time, said, 'Oh, papa! Look! The horses have run away, and there's the carriage running after them! Isn't it funny!'[5] Similarly, the 1907 diary of Dr Tracey described his experiences as the first to run a motor vehicle in his Somerset village.[6]

Motor cars were so novel that there was no established name for them. Sturmey, once underway on his epic trip, went to put his Daimler on a ferry, only to find the ferryman at a loss to match it against his list of tariffs.[7] The new motoring magazines, springing up in the 1890s, discussed what the new 'motor-carriages' should be called; the very titles of *The Autocar* (founded 1895) and the *Automotor Journal* (founded 1896) show two terms then in use. The latter magazine made a list of suggestions in 1896 which did not even include 'motor car' – although that term was probably used as early as 1891,[8] it did not catch on until later. Instead, the magazine thought of, amongst others, 'movers', 'autokinons', 'motes' and 'go cars'.[9] Even the activity of motoring did not have an agreed term: in a letter to *The Autocar* in 1900, 'Oilman' described himself as the 'owner of an "autocar"', but 'I cannot get used to call myself a "motist" yet'.[10]

These carefully selected moments are intended to illustrate just how niche, hobbyist, even reckless the sport of motoring was understood to be by the wider population. In the popular imagination, the motor car was dangerous and likely to blow up. The pioneer motorist Charles Jarrott (1877–1944) observed this when he took a motor car from Margate to London in 1897. The starting ritual involved creating under the bonnet 'rather a big blaze' with petrol just to get the ignition burners lit, a necessary prelude to attempting to turn the engine over by starter handle. Not surprisingly, the onlookers were highly agitated.[11] The very term 'petrol' was novel, referring to a product then available by the fluid ounce and usually used as a cleaning agent. Another early motorist J. A. Koosen (*c.* 1860–1913) recalled, 'My experiences as a pioneer – well, they were simply awful. For a long time I could get no petrol; nobody knew what it was. I then asked the chemists for "benzin", and one of them had some in stock and asked did I want a two-ounce or a four-ounce bottle! When I said something about five or ten gallons he nearly had a fit.'[12]

---

[4] For traction prior to the light motor car see T. R. Nicholson, *The Birth of the British Motor Car, 1769–1897, Vol. 2: Revival and Defeat, 1842–93* (London: Palgrave Macmillan, 1982). See also L. T. C. Rolt, *The Horseless Carriage* (London: Constable, 1950), chap. 1.
[5] *Punch*, 15 October 1902, p. 257.
[6] Hugh Tracey, *Father's First Car* (London: Routledge & Kegan Paul, 1966).
[7] Esme Coulbert, 'Perspectives on the road: Narratives of motoring in Britain 1896–1930' (unpub. PhD dissertation, Nottingham Trent University, 2013), p. 7.
[8] Montagu of Beaulieu and David Burgess-Wise, *A Daimler Century: The Full History of Britain's Oldest Car Maker* (Sparkford: Patrick Stephens, 1995), p. 13.
[9] *Automotor and Horseless Vehicle Journal*, 16 December 1896, pp. 112–13.
[10] *The Autocar*, 17 February 1900, p. 167.
[11] Charles Jarrott, *Ten Years of Motors and Motor Racing* (London: Motor Sport, 1929), pp. 13–18.
[12] J. A. Koosen in [Lord] Montagu, *The First Ten Years of Automobilism, 1896–1906* (London: *The Car Illustrated*, 1906), p. 22. Useful surveys on the first availability of petrol are Edward Liveing, *Pioneers of Petrol: A Centenary History of Carless, Capel and Leonard, 1859–1959* (London: H. F. and

Punch magazine's 1900 'Roll of Fame' was a cavalcade of all that was modern and credible in science and industry. It celebrated the bicycle and the X-Ray, but featured no motor cars. For some people, the motor cars they saw appeared to be uncontrollable: *Punch* in 1901 published a cartoon in which the hapless owner of the 'violently palpitating' motor car attempted to restrain it with a pitchfork as villagers looked on.[13] Even as late as 1905 P. G. Wodehouse was publishing stories in which control of the motor car remained a dark art.[14] For others, the motor car was somehow capable of ridiculous feats. The magazine *The Strand* published stories where motor cars were capable of heroic if implausible mercy dashes.[15] The motor car also featured in the same magazines as a likely sporting replacement for the animal. 'The newest twentieth-century game', was, for example, according to the *Harmsworth London Magazine*, 'motor polo', where 'nimble little racing motor cars replace the trained pony',[16] while in another tall tale in *The Strand*, the wealthy American Mr Hanks impressed a marquesa by bringing his 10hp Daimler into the bull ring. (After his suitable display of heroics, they married.)[17] Automotive company promoters took advantage of such gullibility among the wider public; an artist's impression of the motoring promoter Edward J. Pennington's (1858–1911) motor-cycle in his company catalogue in 1896 featured it leaping over a ravine.[18] There were no trusted brands of motor car, and so companies with other specialisms launching into motor-car production did not necessarily imbue the consumer with confidence. For example, the French company Panhard and Levassor had been a manufacturer of woodworking equipment. The German Daimler company was a manufacturer of internal-combustion engines, which were more likely to find uses as stationary engines in factories, or perhaps to power a boat, than power a motor car.[19]

Finding the open road

Cycling had not long before experienced a similar standing among the wider public. In the 1870s – in the form of the 'ordinary', or 'penny-farthing' – it had been a niche sport

---

G. Witherby Ltd, 1959); and Stephen Howarth, *A Century in Oil: The 'Shell' Transport and Trading Company, 1897–1987* (London: Weidenfeld and Nicholson, 1997).

[13] For example, 'There's no need to be alarmed', *Punch*, 23 January 1901, p. 63.
[14] P. G. Wodehouse, 'The lost bowlers (a cricket story)', *The Strand*, 30 (1905), pp. 298–303. Here, some friends accidentally start a motor car by touching the lever, which 'sets the thing going', and are unable to stop it.
[15] For example, L. T. Meade and Robert Eustace, 'The Sorceress of the Strand', *The Strand*, 25 (1903), pp. 198–212.
[16] [Henry Leach], 'Motor Polo: A fast and furious game, rendered more exciting by the use of automobiles in the place of ponies', *Harmsworth London Magazine*, 10 (1903), pp. 195–6.
[17] R. B. Townshend, 'A motor in the bull ring', *The Strand*, 26 (1903), pp. 575–80. The bicycle had long since made its appearance in the bull ring: see, for example, 'The progress of velocipeding', *The Mercury* [Tasmania], 12 November 1869, reporting on a circus at Nîmes. My thanks to Rosemary Sharples for this reference.
[18] 'Flying Pennington Cycle', in the Pennington catalogue, 1896, image reprod. in H. O. Duncan, *The World on Wheels* (Paris: privately pub., 1926), ii, p. 588.
[19] Montagu and Burgess-Wise, *A Daimler Century*, pp. 9–16. I am grateful to Prof Chip McGoun for making available to me his unpublished paper, 'Automobile commerce and competition in the nineteenth century', delivered at the fourth Annual Michael Argetsinger Symposium, Watkins Glen, 9 November 2018.

for athletic men, and while it was cutting edge in its adoption of the latest technology, when seen on the public highway it had been associated with middle-class arrogance and privilege. Cyclists then took to the roads, usually unsealed and dreadful out of towns, and absorbed abuse and missiles hurled at them by a large majority of onlookers who did not share their enthusiasm. Encroachment on the highway by cyclists and, later, motorists was real; as Denning has found, they 'initiated a contest for the use of public space, challenging centuries of practice in which roads and streets were a public amenity meant for common use'.[20] But by the 1890s, cycling was booming, and among the middle classes was now widely indulged as a fashionable leisure pursuit. The appearance of the tricycle and the tandem from the 1880s had facilitated cycling for ladies and couples. Fashion and general-interest magazines had picked up on the visibility of cycling, and members of the royal family were endorsing the activity. Cycling magazines had sprung up to cater for a wide cross-class interest; cycle racing, where spectators paid to watch, and long-distance record breaking, were reported widely. Cycling technology – in the form of the 'safety' bicycle, with its diamond-shaped frame – was now seen by consumers as, in a sense, mature, and cycling was now perceived as reliable and accessible.[21] While cycling continued to challenge many people's values – the 'scorching' (reckless cycling), going out for a ride on the Sabbath, unchaperoned female cyclists, some wearing 'rational' dress (divided skirts)[22] – by the turn of the century, it had become an acceptable part of life. Cycling had offered something entirely novel: speed, and the freedom of the open road. Until the 1890s, the bicycle had been the fastest device on the public highway. The 'ordinary' by the 1870s was already averaging about 10mph despite the appalling roads. By 1880 crowds were paying to watch cycling 'cracks' cover twenty miles in an hour.[23]

By the First World War, motor traction was to experience a similar shift as it became more widely tolerated and more attractive for consumers, offering increased reliability and ease of use. Motoring adopted technology developed by the cycling industry – the spoked wheels, tubular frames, chains – and utilized the new, light petrol engines such as Daimler's. The potential that motor traction then offered for even higher speeds than cycling on the open road was, for many, bewildering. Writing in 1900, 'The Deserter' said:

> We live in an age of ever-hastening activity and unceasing rush. The motor-car is generally regarded as the embodiment in metal of this characteristic of the century – a monster that goes throbbing through quiet villages and snorting through busy streets with the impartiality of the plague . . . even in connection with the

---

[20] Andrew Denning, 'Transports of Speed', in Michael Saler (ed.), *The Fin-de-Siècle World* (London: Routledge, 2015), p. 386.
[21] See, for example, Carlton Reid, *Roads Were Not Built for Cars: How Cyclists Were the First to Push for Good Roads and Became the Pioneers of Motoring* (Newcastle on Tyne: Front Page Creations, 2014); or Andrew Ritchie, *Early Bicycles and the Quest for Speed: A History, 1868–1903* (Jefferson, NC: McFarland, 2nd ed., 2018).
[22] See Kat Jungnickel, *Bikes and Bloomers: Victorian Women Inventors and their Extraordinary Cycle Wear* (London: Goldsmiths Press, 2018); F. J. Erskine, *Lady Cycling: What to Wear and How to Ride* (1897; London: British Library Publishing, 2014).
[23] Ritchie, *Early Bicycles and the Quest for Speed*, esp. chap. 2, at pp. 62–4; and chap. 4, at pp. 130–1.

Great Trial [the Thousand Mile Trial of 1900, described later], which was to test the vehicle and not record its speed, chronicles of a mile in four minutes have been told in private conversation. These things distress the easy-going man with a respect for the law – and a wife and family.[24]

By 1899 the Belgian racing driver Camille Jenatzy (1868–1913) had driven a motor car at more than 100km/h, and stories appeared in middle-class fiction-based magazines such as *The Strand* of motor cars capable of racing trains. By 1900, even the bicycle, paced by a train, was reported to reach 'a mile a minute'.[25] Duffy has suggested that, with the motor car, speed was 'repackaged as a sensation and a pleasure'. To understand the first few days of cycling and then motoring we need now to

> recapture the excitement of those who drove the first cars or saw one raise the dust on a village street, for whom twenty-five miles an hour was intensely fast. For a brief moment, roughly the first quarter of the twentieth century, the thrill of velocity at any speed was vividly palpable.[26]

Speed and its attendant dangers preoccupied all road users. Literature on (horse-) driving, cycling and motoring often featured witnessed or fictionalized accounts of carelessness or malicious intent by speeding drivers. The travel writer J. J. Hissey (1847–1921), who had been writing on travel since the 1880s and therefore had the perspective of both horse- and motor-travel, wrote in 1906,

> there is a joy in speed, and poetry in it, and danger in it too. But a rush at full speed in a motor car over a lonely road, and a deserted country, wide and open, is an experience to be ever afterwards remembered. Truly, for such a moment, life *is* worth living, and optimism is rampant.[27]

The Victorian middle classes had long sought the 'open road'. Hissey had described his adventure by phaeton into Kent, Sussex and Surrey in the 1880s, bemoaning the incursion of the railway as he came across it.[28] However, it was the railway that enabled the cyclist to get even further afield – Wales, the Lake District, Scotland. Now, the road was the means to discover a new England (and Wales, and Scotland), and suddenly the road – once adequate for its purpose, the train having eliminated much road traffic – was found to be wanting. Cyclists reported in magazines the 'dangerous' bends, cambers and hills which nobody had noticed before. The coming of the 'safety' cyclist, in much greater numbers than the 'ordinary' cyclist, simply accelerated road use, and

---

[24] *Motor Car Journal*, 19 May 1900, pp. 193–4.
[25] Frederick A. Talbot, 'Cycling at a mile a minute', *The Strand*, 19 (1900), pp. 291–6.
[26] Enda Duffy, *The Speed Handbook: Velocity, Pleasure, Modernism* (Durham, NC: Duke University Press, 2009), p. 5.
[27] John Hissey, *Untravelled England* (London: Macdonald, 1906), p. 427, cited in Esme Coulbert, '"The romance of the road": Narratives of motoring in England, 1896–1930', in Benjamin Colbert (ed.), *Travel Writing and Tourism in Britain and Ireland* (Basingstoke: Palgrave Macmillan, 2012), p. 206.
[28] J. J. Hissey, *A Holiday on the Road: An Artist's Wanderings in Kent, Sussex and Surrey* (London: Richard Bentley and Son, 1887).

then the coming of the motorist tipped the balance altogether. The Brighton road, for example, mostly single-track, unsealed, 'thick with white dust during the summer, and very muddy in the winter',[29] might have tolerated countless cyclists on a sunny weekend in the 1890s, but not countless motorists as well.

This book will suggest that there was, like with cycling, a clear shift in the motor-traction movement from a niche, sporting, clubby activity to one with a wider appeal. Improving reliability, a standardization of design, and a second-hand market in the period to 1914 all facilitated this. It will suggest there was a move from the 'old brigade' of pioneering motorists, for whom the journey was all, to a 'steady and careful artisan', who as a consumer was feeding in to a process of incremental product improvement and starting to demand a more reliable motor car or motor-cycle requiring less intervention. The 1880s and 1890s were already a melting pot of private mobility technologies, with the myriad dead-ends of bicycle design and the appearance of the motor-cycle and light motor car.[30] 'Ordinary' cyclists, the very early motorists, horse-driving amateurs, the first 'safety' riders: they all drew on cultures in place to create and join their clubs, to set records and improve the technology, to start up magazines, to find strength and comfort in their numbers. The club offered access to information and the potential to network. However, the exclusivity and certainties of the club changed in the late 1890s and early 1900s, when demand for the 'safety' brought into its orbit cyclists with no interest in the cultures of competition or forming clubs. Not as if most were clubbable anyway; clubs continued to make sure only the right sort were admitted. The slow but inexorable appearance of the owner-driver of the used motor car or motor-cycle in the ten or so years before the First World War reinforced this; it was this 'class' of motorist (and cyclist), then, who by the First World War typified the wider motoring and cycling experience. Studies of registration records for the period to 1914 have demonstrated the transfer of motor cars and motor-cycles from addresses in wealthier areas, to middle- and even working-class suburbs.[31] These new users, while mostly middle class but hardly the clubbable sort, were buying second-hand and buying (or borrowing) maps and guide books when they drove to find the 'open road' for a day at the weekend.

There was a constant interchange of cultural practice and personnel between the different modes of mobility. It will be shown that many motorists, particularly the pioneering ones, came from a cycling background, and often remained cyclists. In addition, though, it was not unusual to see a mixing of modes to suit sporting or leisure interests. Horse riders, for example, accompanied cyclists on club runs, while

---

[29] Selwyn Francis Edge, *My Motoring Reminiscences* (London: G.T. Foulis, 1934), p. 206.
[30] See Paul Smethurst, *The Bicycle: Towards a Global History* (London: Palgrave Macmillan, 2014), chap. 1; Gijs Mom, *Atlantic Automobilism: The Emergence and Persistence of the Car, 1895–1940* (New York: Berghahn, 2015), chap. 2.
[31] See Craig Horner, *The Cheshire Motor Vehicle Registrations, Vol. 1: 1904–07* (Liverpool: Record Society of Lancashire and Cheshire, 2019), which provides abundant evidence of the passing on of vehicles into working-class areas. Vols 2 and 3, to complete the period to 1914, are in preparation. Other recent surveys include Ian Hicks (ed.), *Early Motor Vehicle Registration in Wiltshire 1903–1914* (Trowbridge: Wiltshire Record Society, 2006); and Peter Barlow and Martin Boothman (eds), *'Conspicuously Marked': Vehicle Registration in Gloucestershire, 1903–13* (Gloucester: Bristol & Gloucestershire Archaeological Society, 2019).

some motorists were also drawn to motor-boating or aviation – this cross-pollination is evident by the way that the publishers of the cycle magazines in the 1880s often branched out with new titles which latched on to these new modes. Emphatically, horse traction did not go away, at least not in the period to the First World War. Most people continued to rely on it and had no desire to see its demise; for all its faults, horse traction worked. Many cyclists and motorists would probably have thought the same.

Entrepreneurs

For motor traction to suggest the promise of the open road, it needed the participation of a pioneering group with entrepreneurial drive. However, to fulfil the promise, motoring had to diversify. It had to become less a sport and more a consumer activity if it was to draw on a much wider body of users, who would not be attracted until they saw that motor traction was demonstrably more reliable and less clubby. To promote motor traction in the face of public hostility required energetic entrepreneurs, many of whom had come from a cycling background. They had absorbed its cultures – of the clubs, the racing, the sport. They were accustomed to abuse or indifference as they indulged their hobby. As the first motor agencies and entrepreneurial businesses were established, they started to promise a product which would be the equal of the horse in terms of carrying capacity and hill-climbing. The publisher Alfred Harmsworth (1865–1922, later Lord Northcliffe) edited a collection of articles on the new sport of 'motors and motor driving' in 1902, and emphasized that a motor car, unlike the horse, did not need resting, would not catch a chill, yet could do far more than ten miles or so – the reasonable limit for a horse – in a day.[32] Where the bicycle promised speed and the open road for the fit, motor traction promised more for less effort.

These entrepreneurs had backgrounds in, say, sales or journalism, and many cases were simply transferring their professional attention from the cycle industry to the motor industry. Sturmey is a case in point – he edited *The Cyclist* magazine from 1877 and was prominent in the Cyclists' Touring Club (CTC, now Cycling UK, set up in 1878 to campaign for better roads and conditions for cyclists). His trip from John O'Groat's was promotional, providing exposure for the Daimler motor car, and copy for his magazine. Cyclists had by necessity adjusted and improved their machines, sometimes at the side of the road, in the rain. Motorists did the same when they bought their first motor cars. The entrepreneur had this same gritty experience, and was instrumental in ensuring that feedback, from those motorists who had learnt so much the hard way, made for a better product. Testimonials, and positive results from the many motor trials of the time, were used in advertising copy.

Selwyn Francis Edge (1868–1940) will stand out in this story. As one of many cycling and motoring entrepreneurs in the late-Victorian and Edwardian period, he was, and remained throughout his life, a cycle-club member. Edge made a name for himself, first as a cycle racer in the 1880s and 1890s, and after 1899 as the sole agent

---

[32] John Scott-Montagu, 'The utility of motor vehicles', in Alfred Harmsworth (ed.), *Motors and Motor Driving* (London: Longmans, Green & Co, 1902), esp. pp. 25, 28–9.

for the new, up-market Napier motor car, a new direction for a company that had specialized in precision engineering.[33] Edge went on, though, to become a household name by winning the international Gordon Bennett Cup motor race in 1902 for Great Britain. His methods of doing business, his projection of his image as a masculine hero in a time of change and challenges to certainties, mean for us that now he can serve as a barometer of the opinions and attitudes of the English, their road-traction movements, their nationalism and their attitudes to foreigners.[34]

Historiography and sources

Approaches to the histories of late-nineteenth-century private 'mobilities' have fundamentally changed in the last generation. The present-day global 'system' of automobility as outlined by Urry[35] is being explained now by an unpicking of the context in which motor traction appeared. Understanding mobilities, and particularly 'automobilities', is now better achieved seen through the cultures of representation and national identities.[36] The diffusion of automobilities is now explained through the contexts of gender, technology, class, consumption and power relations (the displacement of the weak, or, the 'making' of the pedestrian).[37] Mom, in particular, in trying to account for the emergence, then the persistence, of the motor car has described the creation of the 'adventure machine' and challenged the 'toy-to-tool' thesis, suggesting that the early days of motoring would be understood, instead, by its participants' 'touring, tinkering, racing'.[38] Older approaches by economic and business histories remain pertinent; Saul, for example, discussed the early motor industry while

---

[33] See Charles Wilson and William Reader, *Men and Machines: A History of D. Napier & Son, Engineers Ltd, 1808–1958* (London: Weidenfeld & Nicolson, 1958); Alan Vessey, *By Precision Into Power: A Bicentennial Record of D. Napier and Son* (Stroud: Tempus, 2007).

[34] For analyses of Edge, see Duncan, *The World on Wheels*, ii, pp. 799–852; G. H. Smith, *Selwyn Francis Edge: The Man and Some of the Things He Has Done* (privately pub., 1928); Craig Horner, 'S. F. Edge: The salad days', *Aspects of Motoring History*, 11 (2015), pp. 43–54; Edge, *My Motoring Reminiscences*; David Thoms, 'Edge, Selwyn Francis (1868–1940)', ODNB, https://doi.org.ezproxy.mmu.ac.uk/1 0.1093/ref:odnb/32970; Malcolm Jeal, 'By and Large: Napier – Britain's first purpose-built racing car in context', *Aspects of Motoring History*, 5 (2009), pp. 36–64; David Venables, *Racing Colours: British Racing Green; Drivers, Cars and Triumphs of British Motor Racing* (Shepperton: Ian Allen, 2008); David Venables, *Napier: The First to Wear the Green* (London: G. T. Foulis, 1998).

[35] John Urry, 'The "system" of automobility', *Theory, Culture and Society*, 21:4–5 (2004), pp. 25–39.

[36] See, for example, Mom, *Atlantic Automobilism*; David Jeremiah, *Representations of British Motoring* (Manchester: Manchester University Press, 2007).

[37] Sean O'Connell, *The Car in British Society: Class, Gender and Motoring, 1896–1939* (Manchester: Manchester University Press, 1998); Barbara Schmucki, 'Against "the eviction of the pedestrian": The Pedestrians' Association and walking practices in urban Britain after World War II', *Radical History Review*, 114 (2012), pp. 113–38; Virginia Scharff, *Taking the Wheel: Women and the Coming of the Motor Age* (New York: The Free Press, 1991); Margaret Walsh, 'Gendering transport history: Retrospect and prospect', *Journal of Transport History*, 23:1 (2002), pp. 1–8; Julie Wosk, *Women and the Machine: Representations from the Spinning Wheel to the Electronic Age* (Baltimore, MD: Johns Hopkins University Press, 2001).

[38] Mom, *Atlantic Automobilism*, esp. chap. 2, 'Racing, touring, tinkering: Constructing the adventure machine (1895–1914/1917)'; Gijs Mom, 'Civilised adventure as a remedy for nervous times: Early automobilism and fin de siècle culture', *History of Technology*, 23 (2001), pp. 157–90.

Harrison looked at the cycle industry.[39] In particular, some recent 'general' histories of motoring, or of famous marques or visionary leaders, some written by amateur motoring historians, are often overlooked by the academic community because they appear to be descriptive or light on referencing,[40] when many draw on a broad and imaginative range of primary sources. This book, then, is probably the first attempt to bridge two schools of history – the academic and cultural historians, and motoring and cycling historians.[41] For cycling history, social, technological and cultural aspects are well represented through the journal of the Veteran-Cycle Club and a regular International Cycling History Conference. Smethurst has recently attempted a global history of cycling.[42] Oddy has pointed out how even now while there are many cultural and political histories of the road, they still tend to focus on either motoring or cycling, not both. He also makes the point that cycling and motoring had effectively diverged, in terms of their mutual interests, by 1906 or so;[43] their commonalities up until then, however, are evident. Studies on, say, the road lobby tend to remain motorist- or cyclist-based rather than engaging the wider collective of road users.[44]

O'Connell's now venerable study of motoring was groundbreaking with its vantage points of class and gender, but despite its title is predominantly concerned with the interwar period.[45] With some exceptions,[46] that is the period where much scholarship

---

[39] S. B. Saul, 'The motor industry in Britain to 1914', *Business History*, 5 (1962), pp. 22–44; A. E. Harrison, 'Joint-stock company flotation in the cycle, motor vehicle and related industries, 1882–1914', *Business History*, 23 (1981), pp. 165–90. See also, for example, Roy Church, *The Rise and Decline of the British Motor Industry: Studies in Economic and Social History* (London: Macmillan, 1994). See also T. R. Nicholson, *The Birth of the British Motor Car, 1769-1897* (London: Palgrave Macmillan, 1982), 3 vols.

[40] For example, Peter Thorold, *The Motoring Age: The Automobile and Britain, 1896-1939* (London: Profile, 2003); Ruth Brandon, *Automobile: How the Car Changed Life* (London: Macmillan, 2002); Steven Parissien, *The Life of the Automobile: A New History of the Motor Car* (London: Atlantic Books, 2013).

[41] Conversely, motoring historians tend to overlook academic studies because they 'look' too dense. For motoring historians see, for example, the work of David Burgess-Wise, Anders Ditlev Clausager, T. R. Nicholson, Jonathan Wood, Malcolm Jeal, Nick Georgano and Michael Worthington Williams. These historians draw on increasingly diverse primary sources and their work is peer reviewed. Three books in particular stand out as excellent works of research: Montagu and Burgess-Wise, *A Daimler Century*; Nick Georgano (ed.), *Britain's Motor Industry* (London: G. T. Foulis and Co., 1995); and Anders Ditlev Clausager, *Wolseley: A Very British Car* (Beaworthy: Herridge and Sons, 2016). More recently the journal of the Society of Automotive Historians in Britain, *Aspects of Motoring History*, has shifted its editorial stance to adopt wider cultural and political histories of motoring.

[42] The club publishes *The Boneshaker* every quarter, while the proceedings of the International Cycling History Conference appear as *Cycle History*; Smethurst, *The Bicycle: Towards a Global History*.

[43] Nicholas Oddy, 'This hill is dangerous', *Technology and Culture*, 56:2 (2015), pp. 335–69 at p. 336.

[44] For example, Ritchie, *Early Bicycles and the Quest for Speed*; Reid, *Roads Were Not Built for Cars*. Also Keith Laybourn with David Taylor, *The Battle for the Roads of Britain: Police, Motorists and the Law, c1890s to 1970s* (London: Palgrave McMillan, 2015); William Plowden, *The Motor Car and Politics in Britain, 1896-1970* (London: Harmondsworth, 1971).

[45] O'Connell, *The Car in British Society*.

[46] For example, Nick Clayton, 'A missed opportunity? Bicycle manufacturing in Manchester 1880–1900', in Derek Brumhead and Terry Wyke (eds), *Moving Manchester: Aspects of the History of Transport in the City and Region since 1700* (Manchester: Lancashire and Cheshire Antiquarian Society, 2004); Craig Horner, 'The emergence of automobility in the United Kingdom', *Transfers*, 2:3 (2012), pp. 56–75; Josh Butt, 'The diffusion of the automobile in the North-West of England, 1896–1939' (unpub. PhD dissertation, Manchester Metropolitan University, 2019).

remains. Law's recent work on road-houses and class in motoring is largely interwar, as is much motor-cycling history.[47] While there has been much written on gender and motoring (and cycling), there is little directly relating to the United Kingdom.[48]

For primary materials, much use has been made of cycling and motoring magazines. In the early period, these magazines reflected a much broader interest between these sports than, say, after the First World War. *Bicycling News*, for example, started in 1876, moved to include motoring in 1900 – changing its name to *Bicycling News and Motor Car Chronicle* – so as to cater better for 'both sections of the wheel world'. To account for the shift, 'The trades and the pastime are necessarily so closely allied that no apology is needed', it said, and 'after all, it is almost impossible to write concerning cycles without touching upon motors, and vice versa'.[49] It started a new regular column by 'a recognised authority on motoring matters', C. W. Brown.[50] The magazine believed that cycling and motoring 'engross the attention of much the same class of people. Granted that motoring appeals to a wealthier *clientèle*; we think it will be found, as it has been found in cycling, that where there's a will for motoring, a way will be found in the majority of cases to provide the money'.[51] Similarly, Temple Press, publishers of *Cycling* magazine which appeared in 1891, launched *Motorcycling and Motoring* in 1902, renamed it *The Motor* within twelve months, and that in turn spawned magazines on commercial vehicles, cyclecars, motor-boats and aviation, all before the First World War. Newspapers, in particular the *Daily Mail*, have been used, while remaining alert to the clear stance of its owner, Alfred Harmsworth, a dedicated pioneer motorist whose editorials endorsed the new worlds of cycling and motoring. The cycling diary and reminiscences of the club-cyclist turned motoring entrepreneur G. H. Smith (1862–1946) are also used.[52] *Punch* magazine, a sixpence satirical weekly at its peak in the late nineteenth century, is used to reflect middle-class opinion and inertia;[53]

---

[47] Michael John Law, 'Charabancs and social class in 1930s Britain', *Journal of Transport History*, 36:1 (2015), pp. 41–57; Michael John Law, 'Driving to the "Super" Roadhouse', *Aspects of Motoring History*, 12 (2016), pp. 49–60; Christopher Thomas Potter, 'An exploration of social and cultural aspects of motorcycling during the interwar period' (unpub. PhD dissertation, Northumbria University, 2007); Steve Koerner, 'The British motor cycle industry during the 1930s', *Journal of Transport History*, 16:1 (1995), pp. 55–76. The online *International Journal of Motorcycle Studies* embraces international sub-cultures of motorcycling and is less concerned with chronological periods.

[48] Georgine Clarsen, *Eat My Dust: Early Women Motorists* (Baltimore, MD: Johns Hopkins University Press, 2008); Marilyn Constanzo, '"One can't shake off the women": Images of sport and gender in *Punch*, 1901–10', *International Journal of the History of Sport*, 19:1 (2002), pp. 31–56; Scharff, *Taking the Wheel*; Wosk, *Women and the Machine*; Sarah Wintle, 'Horses, bikes and automobiles: New Woman on the move', in Angelique Richardson and Chris Willis (eds), *The New Woman in Fiction and Fact: Fin de Siècle Feminisms* (London: Palgrave, 2002), pp. 66–78.

[49] *Bicycling News and Motor Car Chronicle*, 14 March 1900, p. 5.

[50] Charles William Brown (b. c 1866) claimed to have driven 2,000 to 3,000 miles prior to 1900. In 1927 he was a councillor for Finchley Urban District Council: Qualification Form, Circle of Nineteenth-Century Motorists, RAC archive, ACQ2/1.

[51] *Bicycling News and Motor Car Chronicle*, 14 March 1900, p. 5.

[52] The unpub. diary (1891–6) of G. H. Smith; and G. H. Smith, *Some Notes About the Anerley B.C.* (privately pub., 1930).

[53] Richard Noakes, 'Representing "A century of inventions": Nineteenth-century technology and Victorian *Punch*', in Louise Henson et al (eds), *Culture and Science in the Nineteenth-Century Media* (Ashgate: Aldershot, 2004), pp. 151–63; Constanzo, '"One can't shake off the women"'; Henry

through its pages it is possible to chart the emergence of the motor car and the reaction of ridicule by *Punch*'s readers, and then observe how the portrayal was 'normalized' within a mere five years or so. According to Duffy, *Punch* 'waded enthusiastically into the world of "scorchers", "driving habits", speed traps, chauffeurs and choleric majors at the wheel'. It was 'eager to jazz up [its] pages and [its] circulations' by inventing a new language for the new 'thrill'.[54] Editors of the motoring press, no doubt sharing some of its readership with *Punch*, did not fail to spot this shift. They saw *Punch* on the cusp, for example, when the *Motor Car Journal* wrote in 1900: '*Punch* appears, judging from the frequency with which automobile illustrations are now appearing in its pages, to be quite a convert to the motor-car and to have fully caught the enthusiasm of motoring.'[55]

Layout of the book

It is the context of Edge and his peers that will form the mainstay of this book. **Chapter 2** considers the context in which motor traction was introduced in the United Kingdom, and the reasons why the country was so slow in setting up a manufacturing presence. Much of the nineteenth-century culture of the club and its rituals were passed on, unchanged, into the cycling and motor-traction movements. Horse traction was fully understood to be imperfect, but 'worked', and very few saw any sense in moving away from it. The club was illustrative of a social-class system which was also evident within the motoring fraternity, seen by, for example, being able to afford the 'right' clothing, or overnight accommodation.

Chapter 3 then attempts to define the cycling and motoring entrepreneur, paying particular attention to Edge and his circle. There was no stereotype, as the entrepreneur was drawn from all social backgrounds. One commonality, though, tended to be the cycling club, which provided a model for social and commercial behaviour. There was also a tendency for the entrepreneur to want to learn all about the product, often done through hard experience. Entrepreneurs were mostly male, but some light is shed on the key activities of women in the promotion of the cycling and motoring worlds.

Chapter 4 shows that the motor trial was a key opportunity for entrepreneurs to place the product in full view of the consumer, while providing copy for the magazines, and awards for manufacturers to shout about in their advertising. This was not without risk for the entrepreneur, though, as the fragile vehicles were very much in the public eye. A case study is made of the 1900 Thousand Mile Trial. Trials became increasingly stringent and of mixed interest to consumers and manufacturers. They were often staged locally through the offices of the local motor club. But when *The Motor* magazine attempted to organize a trial, an analysis of the furore this generated will show that this particular magazine better understood its place in appealing to a

---

Miller, 'The problem with *Punch*', *Historical Research*, 82:216 (2009), pp. 285–302; R. G. G. Price, *A History of Punch* (Collins: London, 1957).
[54] Duffy, *The Speed Handbook*, p. 217.
[55] *Motor Car Journal*, 19 May 1900, p. 283.

more 'modest' motorist than rivals such as, say, *The Autocar*, or the Automobile Club of Great Britain and Ireland.

Chapter 5 shows how the intending consumer of the new motor traction played a proactive role in guiding the development of the product. As a consequence, motor cars and motor-cycles moved from being understood as niche, unstandardized and unreliable, because this did not bring confidence in the product. By about 1905, though, a standardization, of mechanicals and 'look', was starting to emerge. Motoring became increasingly marketed on the appeal of 'this year's model', and the pioneer of the earliest days became increasingly irrelevant, although the exclusivity of 'old timer' clubs showed just how important the pioneer had been in the development of motor traction.

Motor traction was, above all, sold as a leisure activity, from the three-month European tour for some (usually the clubbable) to the weekend drive or camping trip for others (everybody else). **Chapter 6**, though, suggests that the tour evolved. The pioneers, such as those driving before 1900 or a bit later, derived pleasure and satisfaction in getting an utterly unreliable vehicle to its destination, with its punctures, breakdowns and abuse from bystanders. The consumer, though, came to expect an entirely different experience by the First World War, demanding reliability, infrastructure for repairs, petrol and hospitality, and assistance in the form of guide books and maps. It is this 'open road' that the entrepreneur sold, with its impossibly idyllic views, picnic spots and delightful camping experience, reached by the trouble-free motor car, 'so easy a woman can drive it'. The downside was the incompatibility of the nineteenth-century road network with the motor car in any numbers, leading to congestion, despoliation and speed traps.

Flicking through any motoring magazine of the period quickly conjures up the rose-tinted vision that writers and entrepreneurs had for the future of motor traction. Illustrators fantasized about empty roads, fast and trouble-free motor-cycles and motor cars, some of which would be flying or floating. Another vision, though, was of clogged streets and flying policemen. **Chapter 7** considers the different futures presented, and the 'spin-offs' such as motor-boating and aviation, copy for which appeared in the motoring press. The envisaged future for the roads network is discussed, while the persistence of the horse is considered. By about 1910 the elements were in place for an increasingly socially diverse motoring experience requiring ever less intervention.

2

Resistance to change

On the evening of Sunday 14 November 1896, the president of the nascent Motor Car Club, Harry Lawson (1852–1925), presided over an impressive dinner at the Metropole Hotel in Brighton. That morning, for the Club's 'Tour to Brighton', a parade of motor cars had set out from London, and now, its socially elite motorists and their guests had gathered to mark the occasion. Lawson had timed the event to coincide with the day the Locomotives on Highways Act came into effect. This Act meant that a new class of vehicle, the 'light locomotive', would no longer be subject to the restrictions that had effectively barred it from the highway. It would no longer need to keep to 2mph in towns and have a crew of three. One of the participants that day, Henry Finch-Hatton, the thirteenth Earl of Winchilsea (1852–1927), had amused the party by ostentatiously tearing up a red flag – it was the symbolism that was important here, as the parading of a warning (red) flag in advance of a locomotive had ceased to be a requirement of the old legislation as long ago as 1878. The day was, for advocates of the motor-traction movement, a moment of celebration, and has become known ever since as the 'Emancipation Run';[1] it remains an annual fixture of the veteran-car community. Lawson said the day had come at last, 'of the great deliverance of our roads and highways from the reign of quadrupeds'.[2] He anticipated the wholesale, and rapid, replacement of horse traction by motor traction.

Yet Lawson's predictions were premature. Motor traction languished in the United Kingdom for another five years or more, and most sales were of imports, mainly from France or Germany, or made in this country but under licence. This had not been the case with the bicycle. For that, a European innovation, the British had been quick to set up a home-grown industry from the 1870s,[3] and indeed, 'invented' the 'safety' bicycle

---

[1] See Malcolm Jeal (ed.), *London to Brighton Run (Centenary)* (Crawley: Consortium, 1996); in particular Michael E. Ware, 'The 1896 Emancipation Run', pp. 13–18. See also the chapter 'Horseless carriages', in Piers Brendon, *The Motoring Century: The Story of the Royal Automobile Club* (London: Bloomsbury/RAC, 1997); H. O. Duncan, *The World On Wheels* (Paris: privately pub., 1926), ii, pp. 707–19.

[2] 'Meet of the motor-cars', *The Times*, 16 November 1896.

[3] See, for example, A. E. Harrison, 'The competitiveness of the British cycle industry, 1890–1914', *Economic History Review*, 22:2 (1969), pp. 287–303; A. E. Harrison, 'Joint-stock company flotation in the cycle, motor-vehicle and related industries, 1882–1914', in Richard Davenport-Hines and Peter Treadwell (eds), *Capital, Entrepreneurs and Profits* (Abingdon: Frank Cass, 1990), pp. 206–32; Roger Lloyd-Jones and M. J. Lewis with the assistance of M. Eason, *Raleigh and the British Bicycle Industry: An Economic and Business History, 1870–1960* (Aldershot: Ashgate, 2000).

in the 1880s.⁴ By the 1890s, the middle classes, male and female, had taken up the 'safety', usually British-built, as a fashionable contrivance. The difference with motor cars was that the French, in particular, had shown a particular aptitude for making and selling them,⁵ and were able to develop and test them on restriction-free roads, and in long-distance road races, which were not allowed in the United Kingdom. Entrepreneurs then, not surprisingly, found it easier to sponsor the sale of imports rather than domestic production.⁶

This chapter will consider why the United Kingdom appeared to be slow to engage with motor traction. Outside of Harry Lawson's bubble, there was no appetite for the new locomotion, even with its new legal freedom. Lawson, as a promoter, had a part to play here. Throughout, horse traction persisted, and consumers and sportspeople held on to their traditions – horse-driving is used as an example. As cyclists, and motorists soon after, had encroached on the public highway to the irritation of its usual users, resistance was violent and sharp. Once out on the open road, the first cyclists and motorists found poor roads and poor accommodation. As sports, horse-driving, cycling and motoring had required the formation of clubs, and those clubs in turn hindered a wider acceptance with their exclusivity and suffocating codes of practice. Class barriers remained in place to inhibit a wider acceptance of cycling and motoring.

At the time Lawson was speaking, very few motor cars were in use in the United Kingdom, and the thirty-three starters (out of fifty-eight entries) on his 'Tour to Brighton' could have constituted as many as half of all motor vehicles then in the country. The Run was meant to be a dignified procession to demonstrate the ability and reliability of the motor cars. The instructions issued by Lawson's Motor Car Club urged participants to ensure their motor cars were in 'thoroughly good clean order', and that they should remember 'motor-cars are on their trial in England, and that any rashness or carelessness might imperil the industry in this country'.⁷ There was no lack of curiosity on the part of the public – as many as half a million lined the roadsides to watch the parade, and tens of thousands of cyclists shadowed the motor vehicles on their journey; Lawson had done his advertising of the event well. Instead, though, it is the chaos of the event and the unreliability of the motor vehicles that most people remembered. The correspondent from *The Times* was unimpressed with the cars, noting how they vibrated visibly, and how the occupants were jeered by the crowd. Many motor cars broke down and it was a 'pitiful remnant', something fewer than half, that made it to Brighton.⁸ *Punch* reported that 'many of the guests of the Motor Club went to Brighton [. . .] by a horseless carriage – supplied by the L. B. and S. C. Railway',⁹ making reference to some drivers whose motor cars were loaded

---

4 See Paul Smethurst, *The Bicycle: Towards a Global History* (London: Palgrave Macmillan, 2014), chap. 1.
5 See James M. Laux, *In First Gear: The French Automobile Industry to 1914* (Liverpool: Liverpool University Press, 1976).
6 St John Nixon, *The Story of the S.M.M.T., 1902–1952* (London: SMMT, 1952), p. 21.
7 Duncan, *The World On Wheels*, ii, pp. 709–13.
8 'Motor-car day', *Daily Mail*, 16 November 1896; 'Motor-car run. Watched by mighty crowds', *Daily Mail*, 16 November 1896; 'Meet of the motor-cars'.
9 *Punch*, 21 November 1896, p. 245.

onto the Brighton train by stealth,[10] otherwise they would never have made it. Charles Jarrott, writing in 1906, remembered the public dismay:

> The effect of the run on the public was curious. They had come to believe that on that identical day a great revolution was going to take place. Horses were to be superceded forthwith, and only the marvellous motor vehicles about which they had read so much in the papers for months previously would be seen upon the road. No one seemed to be very clear as to how this extraordinary change was to take place suddenly; nevertheless, there was the idea that the change was to be a rapid one. But after the procession to Brighton everybody, including even horse dealers and saddlers, relapsed into placid contentment, and felt secure that the good old-fashioned animal used by our forefathers was in no danger of being displaced. It was, however, the beginning of the movement, and the start in England of the great modern era of mechanical traction on the road.[11]

Late-Victorian Britain and the motor industry

Nicholson has charted the 'birth of the British motor car' back to steam-powered road vehicles in the 1830s, particularly the steam coaches of Goldsworthy Gurney (1793–1875);[12] these offered a public-transport service within parts of London and as far as Bath. An 1861 Act briefly legalized steam vehicles but even then, 'the private steam carriage had appealed to only a tiny minority'.[13] The Locomotive Act of 1865 was a reaction to the wide belief at that time, in part true, that the monstrously large steam-powered juggernauts were prone to exploding, destroying the roads and emitting too much smoke. With the legislation still in force in the 1890s, this was what had the unintended effect of stunting the development of a domestic motor industry, as the motor cars of the 1890s were subject to the same restrictions. Consequently, any testing on the public highway by inventors and manufacturers was curtailed. Nicholson's account, stopping after the 'Emancipation' Act of 1896, uses the language of 'battle' and 'defeat' (the subtitles of the volumes) and thus betrays a collective hindsight, as the motor car indeed ultimately 'won'.[14]

The example of the 'Baronet' steam-driven 'motor tricycle' is illustrative of how the legislation of the 1860s could be so inhibiting for intending motor-traction manufacturers and motorists later in the century. The Baronet, weighing just 2cwt, and with pedals just like a cycle, was shown by Arthur Bateman (c. 1840–1909) at the Stanley Cycle Club show in 1881, and was the first such exhibit of its kind. However, its designer Sir Thomas Parkyns (c. 1820–95) was subsequently prosecuted under the law as it then applied to all road traction. Parkyns had been driving his tricycle at

---

[10] See also Ware, 'The 1896 Emancipation Run', p. 15; Duncan, *The World On Wheels*, ii, p. 715.
[11] Charles Jarrott, *Ten Years of Motors and Motor Racing* (London: Motor Sport, 1929), pp. 2–5, at p. 5.
[12] See T. R. Nicholson, *The Birth of the British Motor Car, Vol. 1: A New Machine, 1769–1842* (London: Palgrave Macmillan, 1982), esp. chaps 2 and 3.
[13] T. R. Nicholson, *The Birth of the British Motor Car, 1769–1897, Vol. 2: Revival and Defeat, 1842–93* (London: Palgrave Macmillan, 1982), pp. 288–9.
[14] Nicholson, *The Birth of the British Motor Car*.

excessive speed (5mph when the limit was 2); for not having three people in charge of the vehicle; and for having tyres less than three inches wide. This led to the 'Parkyns -v- Priest' case in the High Court, which Parkyns lost. Thereafter, according to Jeal, other budding constructors at the time carried on their development work by stealth.[15]

Yet, even the 'emancipated' post-1896 industry remained stifled, with observers then, and historians since, considering the reasons for this.[16] Recounting the second London to Brighton run in 1897, Jarrott noted that some motor companies which had seemed buoyant at the time of the 'Emancipation run' the year before were now dead or dying. While this underlined a general belief that motors were absolutely impracticable,[17] it also suggested there were other reasons for the stagnation of the British motor industry, which remained sleepy even after legislation had been repealed.

One explanation was cultural. An English translation of the author and critic Max Nordau's *Degeneration* (1894) had just appeared, which saw a 'dusk of nations' driven by a decadence amongst writers, poets, dramatists, artists and composers of the time.[18] Wiener has described a process of gentrification that damaged the British enterprise culture, particularly in the context of rising American and German competition.[19] Leaders may have felt a sense of unease at the increasingly visible challenging of certainties: gender roles being subverted by the shocking militant methods of suffragettes in seeking female representation; the wearing of 'rational' dress; an increased mobility for women by bicycle. Movements such as 'Arts and Crafts' resisted a culture of mass production. Yet, other historians have suggested that Victorian decline was relative, and that, indeed, manufacturing, technology and industry all increased through the Victorian period.[20]

Even Harry Lawson had his part to play in the stagnation of the motor industry. His vision was calculated to make him very wealthy and powerful. He had established the British Motor Syndicate in 1895, the purpose of which was to acquire as many patents as possible pertaining to motor vehicle design and construction. The idea

---

[15] Nicholson, *The Birth of the British Motor Car*, ii, pp. 284–6; Malcolm Jeal, 'Pioneer British motor car makers', in Jeal (ed.), *London to Brighton Run (Centenary)*, pp. 59–63, esp. pp. 59–60.

[16] See, for example, Horace Wyatt, *Common Commodities and Industries: The Motor Industry* (London: Sir Isaac Pitman, n.d. [c. 1920]); Martin Adeney, *The Motor Makers* (London: Harper Collins, 1988), chaps. 1–2; Roy Church, *The Rise and Decline of the British Motor Industry: Studies in Economic and Social History* (London: Macmillan, 1994), chap. 1; Nick Georgano (ed.), *Britain's Motor Industry: The First Hundred Years* (Sparkford: G. T. Foulis, 1995), chaps. 1–3; James Foreman-Peck, Sue Bowden and Alan McKinlay, *The British Motor Industry* (Manchester: Manchester University Press, 1995), intro. and chap. 1. See also William Plowden, *The Motor Car and Politics in Britain, 1896–1970* (London: Harmondsworth, 1971); and Bradley J. Beaven, 'The growth and significance of the Coventry car component industry, 1895–1939' (unpub. PhD dissertation, De Montfort University, 1995).

[17] Jarrott, *Ten Years of Motors and Motor Racing*, p. 23.

[18] Max Nordau, *Degeneration* (New York: D. Appleton & Co., 1895). See Nicholas Freeman, *1895: Drama, Disaster, and Disgrace in Late Victorian Britain* (Edinburgh: Edinburgh University Press, 2011); Sally Ledger and Roger Luckhurst (eds), *The Fin de Siècle: A Reader in Cultural History, c. 1880–1900* (Oxford: Oxford University Press, 2000), pp. xiii–xv.

[19] Martin Wiener, *English Culture and the Decline of the Industrial Spirit, 1850–1980* (Cambridge: Cambridge University Press, 2nd ed., 2004).

[20] See W. D. Rubinstein, *Capitalism, Culture and Decline in Britain, 1750–1900* (London: Routledge, 1990); David Edgerton, *Science, Technology and British Industrial 'Decline', 1870–1970* (Cambridge: Cambridge University Press, 1996).

was that every manufacturer, owner and importer would be liable to his patent royalties. Over the next few years Lawson advertised widely to threaten action against anyone not paying royalties. Injunctions were brought, and this kept investors in the industry – and buyers – at bay. Meanwhile, and as Nixon put it, 'The headquarters of the Syndicate swiftly became a kind of Tom Tiddler's ground for every charlatan in Europe', as all comers approached him to sell their crackpot ideas.[21] Had Lawson been more modest in his scheme, he might have succeeded. Instead, and as an example, the patents he held for the Maybach float-feed carburettor were used to demand a fee for any vehicle whose carburettor was based on that principle. That fee amounted to 10 per cent of the cost of the chassis, and other patents could well be applicable for the same vehicle.[22] Consequently, the Automobile Mutual Protection Association Ltd was established to fight his patent monopoly. In 1901 it agreed to support the motor agent Charles Friswell (1871–1926), who since 1896 had imported Peugeots, a brand liable to many of Lawson's patents. Friswell won, and Lawson's dreams of a monopoly rapidly unravelled.

An apologist for Lawson, the cycling entrepreneur H. O. Duncan (1862–1945), pointed out that Lawson alone was the 'one prominent individual who created and built up from zero the British motor industry in 1895–6', and had been feeding tens of thousands of pounds of capital into the industry.[23] Yet, the pro-motoring *Daily Mail* wrote in 1899:

> there has been a strange dilatoriness in the adoption of motor traction in this country. The almost prohibitive cost of the vehicles, due in considerable part to the greediness of self-seeking financiers had a good deal to do with the unwillingness of a naturally conservative public to take up new ideas, and the ridiculous fiasco of the notorious ride to Brighton [the 1896 'Emancipation Run'] made many persons cautious who would otherwise have been ready to avail themselves of newer methods of road traction. The result is that Great Britain has fallen greatly behind in the manufacture and employment of road motors.

British sluggishness in its motor industry was, in sum, due to 'Over confidence, too much company promoting, and too much legislation'.[24] The practice of huge company flotation was a feature of the late-nineteenth century, taking advantage of a wider interest in the new and glamorous technologies of the bicycle and motor car. Along with Lawson, two other 'self-seeking financiers' were of particular repute: Ernest Terah Hooley (1859–1947) and the American Edward J. Pennington (1858–1911).[25] They took advantage of an absence of regulatory processes in their promotion of share prospectuses and flotations. Harrison has enumerated public-flotation activity in the cycling and motoring arenas and showed how the peaks of activity coincided with their scams in 1896 and 1897 and a 'safety' bicycle boom. Payne has described the

---

[21] Nixon, *The Story of the S.M.M.T.*, pp. 18–19.
[22] Nixon, *The Story of the S.M.M.T.*, pp. 19–20.
[23] Duncan, *The World On Wheels*, ii, pp. 672–4.
[24] 'The coming craze. Why we have been beaten. Can we recover?', *Daily Mail*, 4 October 1899.
[25] Duncan, *The World On Wheels*, ii, part XII, pp. 763–95.

widespread practice of promoters underwriting shares, of the 'beguiling and colourfully printed' prospectuses to lure investors, of the practice of buying space in newspapers for 'puffs' and 'write-ups', and of bribes paid to journalists.[26] The techniques of the promoters were audacious; for example, Hooley rented a floor of the Midland Grand Hotel at St Pancras, London, to impress visitors, and had instructed his typewriters (as typists were known) to type anything as visitors passed through, to give an impression of intense business activity.[27] The cycling, motoring and national press all cautioned against investing in any of their schemes.

Furthermore, Duncan had suggested that British companies wasted much of their capital on developing engineering solutions to problems that foreign manufacturers had already worked through. As Duncan put it, 'the one important feature to secure immediate success was a British motor-engineer who would condescend to copy a foreign-made car. This is an incontrovertible fact.' Instead, 'the truth is that no one at that period was anxious to copy a foreign-made machine on the principle that "England was first and foremost in mechanical skill"'.[28]

National pride, then, was a factor. Frequently, letters to the editors of the motoring magazines bemoaned the absence of a home industry of any size or quality. The *Daily Mail* hoped that the 'comparative fiasco' of the 'Emancipation Day' Run in 1896 would bring home just how much the United Kingdom was lagging behind France.[29] Even by as late as 1905, the pioneer motorist and journalist Major C. G. Matson (1859–1914)[30] was writing in the same newspaper how at that year's Paris motor show, there was still a tendency by the buying public towards the cheaper car, and here Britain took half of all French output:

> It is all very well to say [he wrote] that 'it is unpatriotic' to use foreign motors when the home article is available, but no reasonable person by any stretch of imagination can candidly affirm that the manufacture of self-propelled carriages in this country is a national industry. We make certainly some splendid cars, but not enough of them. Many firms started on hopelessly wrong lines, and have had to alter, even quite recently, all their designs, and some vehicles, of which

---

[26] A. E. Harrison, 'Joint-stock company flotation in the cycle, motor vehicle and related industries, 1882–1914', *Business History*, 23 (1981), pp. 165–90; P. L. Payne, 'The emergence of the large-scale company in Great Britain, 1870–1914', *Economic History Review*, 2nd ser., 20 (1967), pp. 519–42, at pp. 522–3. See also A. E. Harrison, 'Growth, entrepreneurship and capital formation in the United Kingdom's cycle and related industries, 1870–1914' (unpub. PhD dissertation, University of York, 1977). More recently, Amini and Toms have suggested that having aristocratic directors was a prerequisite for access to capital markets, rather than being intended to impress: Shima Amini and Steven Toms, 'Accessing capital markets: Aristocrats and new share issues in the British bicycle boom of the 1890s', *Business History*, 60:2 (2018), pp. 231–56.

[27] See E. T. Hooley, *Hooley's Confessions* (London: Simpkin, Marshall, Hamilton and Kent, n.d. [1924]).

[28] Duncan, *The World On Wheels*, ii, pp. 673, 734.

[29] 'Motor-car day', *Daily Mail*, 16 November 1896.

[30] Charles George Matson (1859–1914) of the Royal Marines Light Infantry, 'permitted to retire at his own request on a gratuity' in 1900: see Roger Hurst, 'Matson family of East Kent' (2009), http://acms.sl.nsw.gov.au/_transcript/2014/D32715/a1453.pdf, p. 116, accessed 28 November 2019. Matson contributed to *The Autocar*, *Daily Mail*, *Badminton Magazine* and *Automotor and Horseless Vehicle Journal*. He also contributed to the *Bystander*.

only a year ago much was heard, are now entirely obsolete and will soon be quite unsaleable.[31]

An article in *The Motor* in 1905, 'British cars for Britishers' called it 'a matter of regret' that continental manufacturers should gain such a hold on us.[32] The early motorist Mary Kennard (c. 1844–1936), writing in *Motoring Illustrated* in 1902, asked why buyers 'obstinately' bought 'continental importations'. With 'many types of splendid home-built cars', she wrote, the 'ignorant and foolish British public' 'continue to patronise foreigners, whose workmanship is frequently gimcrack in the extreme, and which cannot compare with the British-made article for wear'. Why not patronize the 'good, honest, reliable workmanship produced in [our] native land?'[33] Mrs Kennard had been an early supporter of the British Napier brand, which was a collaboration between the 'ordinary'-riding cyclist Montague Napier (1870–1931) and its sole agent S. F. Edge. Napier himself related in the *Automotor Journal* the danger of losing business 'as long as France was allowed to overshadow England in the design and construction of automobiles'.[34] Edge wrote to *The Times* to point out that Napier was the only British brand actively fighting off the foreigners[35] and certainly, by racing Napier cars, Edge was one of the few brave, or brazen, enough to represent his country by campaigning a British product in open competition. In floating S. F. Edge (1907) Ltd, he used as a strapline in his advertising: 'protect home industries by producing at home an article so good that the foreigner cannot compete with it'.[36] Yet, Edge's interests included the agencies for many foreign brands, including Gladiator and Clement-Panhard. Many of Edge's peers did the same: Jarrott, for example, acted for De Dietrich and Oldsmobile, while Duncan, Edge and Jarrott combined in 1899 to form the De Dion British and Colonial Syndicate, to control all domestic sales of the French De Dion brand.[37]

A near-xenophobic spirit prevailed amongst all classes and was articulated in the press. 'Wake up, England!' bemoaned an editorial in *The Motor* in 1904.[38] Jarrott related an account of speeding in Ireland in preparation for the 1903 Gordon Bennett Cup race. He was stopped by a policeman who, realizing Jarrott was one of the racers, urged him to beat 'them furriners', especially 'them Germans'.[39] *Punch* was particularly sensitive on this matter: a 1903 cartoon showed a Lanchester (that is, a quintessentially British) motor car whose occupants did not expect to be prosecuted for speeding 'in our sober, respectable-looking car. [. . .] Of course, if you go about in a blatant, brass-bound, scarlet-padded, snorting, foreign affair [. . .] you are bound to be dropped

---

[31] *Daily Mail*, 21 December 1905.
[32] *The Motor*, 18 April 1905, p. 288. Clausager also discusses the spate of articles in *The Times* in 1905 urging motorists to buy British rather than French: Anders Ditlev Clausager, *Wolseley: A Very British Car* (Beaworthy: Herridge and Sons, 2016), p. 79.
[33] Mary Kennard, 'Why buy foreign cars', *Motoring Illustrated*, 22 March 1902, p. 66.
[34] *Automotor and Horseless Vehicle Journal*, 20 November 1909, p. 1393.
[35] *The Times*, 9 June 1908.
[36] 'Motor matters', *The Times*, 11 July 1907.
[37] Duncan, *The World On Wheels*, ii, p. 759.
[38] *The Motor*, 6 April 1904, p. 245.
[39] Jarrott, *Ten Years of Motors and Motor Racing*, p. 264.

on, no matter how slow you go'.[40] The press generally talked up the domestic product and welcomed contributions from industry pundits such as Edge, even if he did tend to use them as advertising 'puff' pieces. Edge, for example, wrote to *The Times* in 1910 condemning smoky exhausts; he was sure that Napier taxicabs didn't smoke, he wrote, but lots of foreign cars do.[41]

Horse traction

The magazine *The Strand* had noted in 1903 how 'it is impossible to take one's stand in Piccadilly or Hyde Park [in an] afternoon and not perceive that a great change is coming over our vehicular traffic'. There are now, it said, 20,000 automobiles, offering employment to nearly 20,000 drivers or chauffeurs. 'One is almost getting accustomed to the auctioneer's advertisement: "Sir Blank Blank is giving up his stables, and in consequence the whole of his stud will be sold on the – inst., without reserve".' The magazine went on to identify the difficulties of then retraining staff for the new motor traction. For example, 'A school for chauffeurs'[42] discussed the resistance by coachmen, grooms and footmen to retrain for a motoring environment, and the despised race of 'motor stokers' and 'cheffoneers' [chauffeurs]. The story then described the London school of Mr Ernest Livet of the Daimler Co., and the coachmen and footmen gathered together for their motoring lessons. One [William] beheld a motor car and said: 'Huh! Looks like a bloomin' iron foundry. My guv'nor'll never get me to chuck the last old crock in the stables for *this*! This ain't natural, this ain't! I'd rather drive a double tandem of mokes[43] any day. They'd have four legs apiece and a tail, anyhow!' The instructor remarked of the rest: 'They'll never make engineers. They may be able to drive through a bit of straight country and clean the motor, but they lack the brains for machinery. They are past thirty, and – well, their hearts are with the horses.'

While the inexorable rise of motor traction is reported in the national press, then, the horse remained utterly dominant in the period to the First World War. The motoring-novelist Alice Williamson (1858–1933) remembered how, in about 1901, 'Automobiles were being talked about, though seldom used, and the new [motor] taxis stood in neglected ranks while habit-bound people shuddered, shrank from the sight of them, and passed on to hail a good old hansom [cab]'.[44] Gordon's *Horse world of London*, published in the 1890s, enumerated the breadth of the scale and investment in horse traction in the capital. Historians since, such as Thompson and Turvey have explored the immersion of the horse in the late-Victorian social landscape, including its management, costs and persistence in the urban environment.[45]

---

[40] *Punch*, 18 February 1903, p. 123.
[41] *The Times*, 27 September 1910.
[42] 'A school for chauffeurs', *The Strand*, 26 (1903), pp. 386–92.
[43] Arranging horses in tandem to pull a carriage was well known to be fraught, while a moke meant a worn-out horse.
[44] Alice Williamson, *The Inky Way* (London: Chapman and Hall, 1931), p. 57.
[45] W. J. Gordon, *The Horse World of London* (London: The Religious Tract Society, 1893); F. M. L. Thompson, 'Nineteenth-century horse sense', *Economic History Review*, 29:1 (1976), pp. 60–81;

In this context, the imposition of the motor car introduced dangers presenting real alarm. For example, there was an association amongst a wider public of petrol with explosion. The entrepreneur Frederick Simms (1863–1944) had been denied a licence in 1891 for exhibiting a Daimler petrol engine at the German Exhibition at Earls Court, because of the perceived danger of petrol. For the same reason Simms was also denied use of a Daimler-driven motor boat as a demonstrator for the potential of the engine on a lake near London, so had to use the Thames. Simms had already anticipated problems with the authorities when he turned down Daimler's offer of a motor car as a demonstrator for the engine, fully aware of the restrictions he was likely to face on the road; that was why he chose a motor boat instead.[46] The first motorists, when on tour, reported many problems 'stabling' their motor cars overnight, as the landlords feared their property might burn down as the cars spontaneously combusted.

It was evident motoring had potential but little to make it a credible replacement for the horse. The cycling 'crack' F. T. Bidlake (1867–1933) had become a journalist and a convert to motor traction by 1900, but his description of the apparent effortlessness in driving a motor car would have made an absurd read for those outside of the motor-traction bubble. That he was also arguing for a rise in the speed limit for motor cars would also have been alarming:

> Shall we, then, suffer to be confined to twelve miles an hour when our horses are abolished, the untiring power of a dozen of the best of them is packed away in our carriages, and that power is obedient to a lady's lightest touch, on a vehicle cushioned from vibration, steerable to a hair's-breadth, and stoppable in a fraction of a second? Emphatically not.[47]

Before cycling, there had been little precedent in using roads for leisure. From the perspective of the cyclist, the roads were poor, narrow and unmaintained, as described in detail by diarists and chroniclers of the first cycling clubs.[48] By contrast, the tourist J. J. Hissey found little about the roads to complain when he explored rural Kent in a phaeton in the 1880s.[49] Nicholson has pointed out that the condition of the road mattered much less for the horse.[50] Instead, guidance offered to horse-driving beginners in the late-nineteenth century concentrated on mastering the team of horses and did not mention road conditions.[51]

Ralph Turvey, 'Horse traction in Victorian London', *Journal of Transport History*, 26:2 (2005), pp. 38–59. For the American experience see Clay McShane, *The Horse in the City: Living Machines in the Nineteenth Century* (Baltimore, MD: Johns Hopkins University Press, 2011).

[46] Nixon, *The Story of the S.M.M.T.*, pp. 17–18. Burgess-Wise believes that Nixon is mistaken but the wide fear of the spontaneous combustion of petrol was evident: Montagu of Beaulieu and David Burgess-Wise, *A Daimler Century: The Full History of Britain's Oldest Car Maker* (Sparkford: Patrick Stephens, 1995), pp. 13–14.

[47] *Bicycling News and Motor Car Chronicle*, 2 May 1900, p. 19.

[48] See, for example, the unpublished diary (1891–6) of G. H. Smith; and G. H. Smith, *Some Notes about the Anerley B.C.* (privately pub., 1930).

[49] J. J. Hissey, *A Holiday on the Road: An Artist's Wanderings in Kent, Sussex and Surrey* (London: Richard Bentley and Son, 1887), chap. 15.

[50] Nicholson, *The Birth of the British Motor Car*, ii, p. 292.

[51] Major Henry Dixon, 'Hints to beginners. Part I', in [His Grace the Duke of] Beaufort, *Driving* (London: Longmans, Green & Co., 4th ed., 1894), pp. 116–30; Colonel Hugh Smith-Baillie, 'Hints

Nicholson has suggested that the private motor car was perceived as a luxury, with 'severe social drawbacks'. He continued:

> It was associated with eccentrics who defied convention and even the law; and in an age of beautiful horsedrawn turnouts, it was usually a massive, inelegant, dirty beast – a vehicle for mechanics to be seen in, not the gentry'.[52]

The horse-drawn coach remained the marker of status and luxurious road travel. Coachbuilders, as late as the First World War, persisted with their established business models in which carriages were bespoke and could last for generations. It was difficult for them to make a sensible business case for accommodating the motor car, even if motor-body purchase was likely to be much more frequent, meaning more business. Cockshoots of Manchester serves as one example of a coachbuilder whose board agreed to shift from carriage- to motor-car-body production in 1903, but with a significant minority (two out of six) resisting the move.[53] The association of the motor car with dirt, unreliability, smoke and danger proved a considerable hindrance for its wider acceptance. Jarrott, writing in 1906 and trying to rationalize why motorists were initially objects of 'ridicule, scorn and abuse', decided 'we English are extraordinary in our conservatism and love of the old order of things. We cling to our old traditions and usages – not because they have any particular merit, but because they are old.'[54] This mentality applied to all technologies: some older cyclists persisted with the 'ordinary' long after the 'safety' had appeared.

The motor car, then, where it cropped up, became an additional mode of mobility, alongside the horse (and bicycle, train, tram, walking and omnibus). Some considered cycling and motor traction as an additional way of getting about, not supplanting but complementing the horse. G. H. Smith (1862–1946) of the Anerley Bicycle Club (ABC) kept a cycling diary in the 1890s and in later years wrote an unofficial history of the club. In it he noted how, on one expedition to Alton, the Letts brothers 'elected to hire horses and have a change from cycling'. The brothers were, apparently, used to horses, so had no problem, but another who joined them, W. H. M. Burgess, had a horse who sensed its rider was inexperienced and returned him to the stable yard.[55] Smith also recorded how the first captain of the ABC, H. S. Hughes, 'relapsed into horse riding and became proprietor of a very horsey paper'.[56]

Despite the juxtaposition of evermore cartoons on a motoring theme, the horse continued to be depicted as the natural means of mobility in the cartoons of *Punch* for the entire pre-First World War period. Morrison and Minnis also show

---

to beginners. Part II', in Beaufort, *Driving*, pp. 131–7.
[52] Nicholson, *The Birth of the British Motor Car*, ii, pp. 288–9.
[53] See G. N. Georgano (ed. in chief), *The Beaulieu Encyclopedia of the Automobile: Coachbuilding* (London: The Stationery Office, 2001), pp. 3–13; Josh Butt, 'Adapting to the emergence of the automobile: A case study of Manchester coachbuilder Joseph Cockshoot and Co., 1896–1939', *Science Museum Group Journal*, 8 (2017), n.p. Butt also emphasizes the close relationship between coachbuilder and customer, and this would naturally spill over into motor-car-buying culture.
[54] Jarrott, *Ten Years of Motors and Motor Racing*, p. 284.
[55] Smith, *Some Notes about the Anerley B.C.*, p. 22.
[56] Smith, *Some Notes about the Anerley B.C.*, p. 8.

photographic evidence of the continued integration of the horse in city traffic after the War,[57] and Butt has suggested that on wider urban thoroughfares, for example, Deansgate in Manchester, an unwritten rule of the road kept horse-drawn (slow) traffic to the nearside lanes and motorized traffic to the 'overtaking' lanes.[58] The horse, then, remained part of the mobility mix.

The persistence of an interdependent world of horsemanship, heritage, skill and tradition was evident. The Badminton Library of Sports and Pastimes published an edition on horse-driving in 1889, edited by the Duke of Beaufort (1824–99) – a clear success as it went into at least four editions.[59] Horse-driving remained a popular choice for amateur sportsmen, and Beaufort's book sheds light on their privileged world, with their codes and uniforms. Yet, while it celebrated a traditional practice, it also described a mid-century 'coaching revival' which was to last into the twentieth century. The contributor W. C. A. Blew described the re-creation of a national network of passenger-carrying (horse-)coaches, all run for profit, and many all year round. These coaches were not (just) patronized by 'heritage' day-trippers, hankering after a long-gone past, but were taken on routes 'in order to meet the convenience of [the] up-passengers, [running] straight to the Royal Exchange in the morning, so as to land City Men at the doors of the places wherein the golden calf had to be worshipped'.[60] As many as twelve coaches ran out of London daily by 1873, some running all year. In addition, and as a measure of coaching as a 'sport', the coaching speed records, such as those by celebrated coachman James Selby (1844–88) with his 'Old Times' coach-and-four, were set in the 1870s and 1880s, long after the 1840s' 'demise' of the coach in the face of competition by the train.[61]

Diaries and accounts remain, describing, to modern readers, an astonishing willingness to tolerate extremes of cold, pain and discomfort when riding and repairing, at the roadside, those first bicycles and motor cars.[62] But horse-driving as a sport offered a context for the apparent hardiness of those very cyclists and motorists. Rather than seeing these cycling and motoring pioneers as knowing they are 'right' to put up with the discomfort, instead, all mobility at this time, the upper-class carriages of the train excepted, was bitterly uncomfortable. What the motoring pioneers were 'enduring' was normal. Beaufort's book on horse-driving describes carriage-riding as so uncomfortable that it was actually better to sit outside than in, even on the coldest

---

[57] Kathryn A. Morrison and John Minnis, *Carscapes: The Motor Car, Architecture and Landscape in England* (New Haven, CT: Yale University Press, 2012), chap. 12, 'Traffic in towns, 1896–1939', esp. pp. 319–22.

[58] Josh Butt, 'The diffusion of the automobile in the North-West of England, 1896–1939' (unpub. PhD dissertation, Manchester Metropolitan University, 2019), p. 152, fig. 26.

[59] Beaufort, *Driving*.

[60] W. C. A. Blew, 'The coaching revival', in Beaufort, *Driving*, pp. 286–8.

[61] For example, and as seen, his London-Brighton-London record of 1888. Curiously, the later editions of the book make no mention of the records since being broken by cyclists, one such being S. F. Edge. Beaufort (as far as I can see) makes no mention of the bicycle, apart from using a picture of 'ordinary' riders scaring the horses: Beaufort, *Driving*, p. 19.

[62] Examples can be found in any motoring pioneer's autobiography, but see, for example, Edge's description of being at the roadside for fourteen hours attending to pneumatic tyres that kept on deflating: Selwyn Francis Edge, *My Motoring Reminiscences* (London: G.T. Foulis, 1934), pp. 209–11.

nights.⁶³ Indeed, his volume reproduces images of those outside passengers, with maybe a dozen people arranged on the roof, raising the centre of gravity and increasing the prospect of the carriage to overturn. It also described the ever-present peril of the runaway or startled horse.⁶⁴ The runaway coach had dreadful potential for injury and death, just as it was for the 'ordinary' rider when striking an obstacle in the road, or getting out of control going down a hill. Cycling, then motoring, brought the same visceral, physical – and terrifying – experience as horse-driving.

The romance of horse-driving remained part of the fabric of society into the twentieth century. Carriages, like locomotives and boats, were given names, and an industry of workers took pride in learning the skills to have command over their 'team' of horses. Plus, in the last resort, with a driver who was ill (or drunk, or asleep) the cart-horse would drive itself home. A measure of the sense of horse-driving having been 'perfected' is when Beaufort discussed the attempt to introduce a drag-brake onto the horse-drawn carriage. This had been met with disdain, in part because it had not been necessary before, but in the main because the driver was, or should be, skilled in reading the road and in conveying instructions to the horses, whose braking effort alone was normally perfectly adequate.⁶⁵ The de-skilling associated with introducing brakes was seen as part of a long-term decline in standards: indeed, where once it took only two minutes or so to change horses at a staging post, it now took five to ten minutes.⁶⁶ Around 1900, cycling underwent a similar 'crisis' of de-skilling, as increasing calls for an effective brake to be fitted to bicycles were met with resistance and disdain from commentators and seasoned riders.⁶⁷

Finally, two small footnotes accompanied the news coverage of the 'Emancipation Run' in 1896. The first was in the *Daily Mail*, which noted that several horse-drawn coaches had gathered with the crowds of onlookers, the intention being to demonstrate that the journey to Brighton 'was still the affair of the horse'. They included the 'well-known coaching guard' Harry Batt. There was a polo pony ready to make the journey, and other horses in place for the relays – this was almost James Selby in his 'Old Times' all over again. 'They were resolved to do or die that day for the honour of sport and the love of the horse', the newspaper reported. The other item appeared in *The Times*, critical of the crowd management. The police, it said, were accustomed to managing crowds at meetings of horse-driving clubs such as the Coaching or the Four-in-Hand Clubs, so that, unlike with the motor cars, 'the vehicles [coaches] would file off, one after the other and be seen by all those who had come to witness the procession'.⁶⁸ The clear message was, the horse-driving world was one that had been mastered, commanding respect and skill while motor traction brought incompetence.

---

63 Beaufort, *Driving*, p. 219.
64 Beaufort, *Driving*, pp. 189–91.
65 Beaufort, *Driving*, pp. 183–5.
66 Beaufort, *Driving*, p. 187.
67 Evident in letters and comments to *Bicycling News and Motor Car Chronicle* in 1900: see, for example, 31 January, p. 29; 7 February, p. 23; 28 February, p. 22; 7 March, p. 23, although the arguments here stemmed mainly from the type of brake used rather than the need for one.
68 'Motor-car run. Watched by mighty crowds'; 'Meet of the motor-cars', *The Times*, 16 November 1896.

Resistance

Plowden has described the animosity felt by the first cyclists and motorists,[69] an experience the pioneer motorist Alfred Harmsworth keenly remembered. His letter was read out at the Royal Automobile Club (RAC) banquet in 1921 to celebrate the twenty-first anniversary of the Thousand Mile Trial:

> By 1900 many of us were already hardened chauffeurs and had toured many parts of England and elsewhere. To us, therefore, the astounding prejudice displayed at that time in England was a matter of amazement and certainly not flattering to our national powers of vision. My newspapers [*Daily Mail*, in particular] received their customary share of hearty abuse for advocating the motor-vehicle. I daresay some present will remember the chorus of diatribes from almost every newspaper against these foreign, ill-smelling nuisances, and how they would ruin the alleged national industry of horse-breeding. They were rich men's toys, calculated to set class against class. The fumes were very injurious, they threw up dust which spread disease.[70]

Punch magazine provided a useful guide to how the middle classes saw this development, with frequent cartoons of put-upon cyclists and motorists. Parodying Mary Kennard's recently published novel *The Motor Maniac* (1902), the magazine published an account, in the style of a novel, of a lunatic asylum where the inmates were all motorists.[71] As a measure of how entrenched the horse-versus-cycle/motor-traction sides could be, in 1900 Mr R. Tweedy Smith, later a Liberal parliamentary candidate, was assaulted by two drunken stablemen as he was stabling his motor car at Putney, telling him to 'get a respectable horse and cart'. A prosecution was brought, for which the assailants were 'let off' with a twenty-shilling fine plus two shillings' costs, 'cheap enough to make the baiting of automobilists a popular movement'.[72] At much the same time 'Cyclomot' wrote in *The Motor* on 'Children and Accidents'. Small boys, he said, were a 'decided nuisance' for drivers. He reported how 'a gang of boys deliberately leapt out in front of the car to cause me to swerve. As usual, I swerve inwards, thwarting them, but this time there was a constable on duty who laughed. Boys who throw caps [at me] are equally a nuisance, and I "collect" them if I can'.[73] Another letter in *The Motor* described the 'small boy tyrant'. It was common for boys to jump out, or throw clothing, mud or stones at motor cars. 'With the existing "popular" feeling against motorists', 'Medicus' said, a motorist will not react if he thinks he will have an unsympathetic crowd gathering.[74] In addition, while cycling and motoring increasingly diverged in the twentieth century, they remained lumped together as a progressive and unwelcome collective: *Punch*

---

[69] See William Plowden, *The Motor Car and Politics in Britain, 1896–1970* (London: Harmondsworth, 1971), chap. 2: 'All the strawberries were spoiled: The motor car and the roads problem'.
[70] *The Times*, 13 May 1921.
[71] *Punch*, 2 September 1903, p. 153.
[72] *Motor Car Journal*, 18 August 1900, p. 406.
[73] *The Motor*, 2 May 1905, p. 348.
[74] Letter from 'Medicus', *The Autocar*, 4 March 1905, p. 335.

published a cartoon in 1908 in which an 'elderly sportsman' on horseback asked a boy on horseback where the 'best inn' was, and the lad replied, 'Well, there's the Wheeler's Arms on the Green where the cyclists and the motor folk goes, but where the gentry mostly goes is to the Fox and 'Ounds up the town.'[75]

The early cyclist or motorist could not rely on 'justice' by the police. Jarrott, for example, was summoned and successfully prosecuted for speeding even though he could prove he was not even driving at the time.[76] It was not unknown for mounted police to run cyclists down.[77] Violence between all parties, threatened and real, was all too apparent. It had been reported how the Marquis of Queensberry had asked a magistrate if he could carry a revolver or rifle to protect himself and his children from sudden death (by a passing motorist).[78] Sir Ralph Frankland-Payne-Gallwey (1848–1916) wrote to *The Times* about wanting to seek the legalization of the use of the shotgun against the motorist.[79] *Punch* wondered 'what it might come to if people carry fire-arms for use against motorists', publishing a cartoon of an armoured car with armoured occupants while a bystander fired a gun at them.[80] Firearms were in relatively wide use. For example, the long-distance cyclist and ABC member G. P. Mills (1867–1945) was reputed to have shot five dogs with his revolver when they bothered him while he was out training.[81] The racing driver Dorothy Levitt (1882–1922) cautioned women 'driv[ing] alone in the highways and byways' to carry a 'small revolver', mentioning the automatic Colt she carried with her, giving her the means to defend herself 'should the occasion arise'.[82] The Edge family also had access to, and wielded, firearms. Edge was fined five pounds in 1893 for threatening to shoot four cyclists, and then twenty shillings for not having a licence for the revolver.[83] Menace seemed to be a family suit: Edge gave a character reference when his younger brother William (1874–1915) appeared in court in 1901 for threatening a man, again with a loaded revolver.[84]

The Autocar, in an article, 'The crusade against motorists', reported on a meeting in Warwick 'to form a county association for the protection of the public against reckless driving of motorists', led by Lord Willoughby, master of the Warwickshire hounds. They need reminding, said *The Autocar*, that the principal menace is, instead, the horse and labourer.[85]

Charles Jarrott remembered the reaction to motor cars:

> Of course, if one stopped for any cause at all, either in the street or in the country, a 'breakdown' was duly recorded. In fact, it was painful to sit on a motor-car, or be

---

[75] *Punch*, 1 January 1908, p. 7.
[76] Jarrott, *Ten Years of Motors and Motor Racing*, pp. 44–6.
[77] E. J. Southcott (ed.), *The First Fifty Years of the Catford Cycling Club* (London: G. T. Foulis, 1939), p. xi.
[78] *The Autocar*, 3 June 1905, p. 735, based on 'The speed of motor-cars', *The Times*, 25 May 1905.
[79] *The Motor*, 10 June 1903, p. 407.
[80] *Punch*, 7 June 1905, p. 397.
[81] David Birchall, 'George Pilkington Mills and the first International Bordeaux-Paris race', *The Boneshaker*, 197 (2015), pp. 12–19, at p. 12.
[82] Dorothy Levitt, *The Woman and Her Car: A Chatty Little Book for Women Who Motor Or Want to Motor* (1909; London: Hugh Evelyn, 1970), p. 30.
[83] *Dundee Courier*, 24 June 1893; *Glasgow Herald*, 30 June 1893.
[84] *The Times*, 23 November 1901; *Daily Mail*, 23 November 1901.
[85] *The Autocar*, 18 March 1905, p. 397.

anywhere near one, whilst it was stationary. Omnibus drivers would hurl at one's unoffending head the advice to 'take it home'; facetious carters would offer the assistance of a horse; and small boys would yell in derision that 'another oil-can had broken down'. At first this tortured my sensitive nature greatly, and I would be overwhelmed with confusion, to cover which I generally dived underneath the car, not because it was necessary, but merely that by so doing I could hide my head. However as time went on familiarity bred contempt. I cannot say that I ever became an expert in retaliation, but in the cultivation of a contemptuous smile I became proficient enough. Not that one did not *feel*; and many a time I would have given much money, and forgone many pleasures, to have been able to expostulate with some of my tormentors in the real old-fashioned British way.[86]

These were violent, physical, times. Edge dealt summary justice to any bystanders foolish enough to throw missiles at him as he cycled or drove by – all cars were open in the earliest days, the occupants exposed to weather and attack.[87] 'In the early days', Jarrott said, 'one always had to be prepared to protect oneself when out motoring, as the sympathy one obtained from the populace and from the authorities was conspicuous by its absence.'[88] As cyclists, Jarrott and Edge had been victims several times. In 1893, then living in Coventry, Edge was cycling with E. Oxborrow (an agent from Rudge cycles) and Mauritz Schulte (1858–1933, from Triumph cycles) when a passing (horse-) driver horsewhipped them. They gave chase but did not catch him.[89] Jarrott referred to such trap-drivers and bystanders as 'Road Hogs', recalling how he 'expostulate[d]' with pedestrians 'in the fashion most approved of by Englishmen'. At one extreme, wire or rope was stretched across the road – 'an old trick on the continent' – intended to injure, even decapitate.[90] Boulders were strewn across the road by 'tramps' to upset a car.[91] Edge and Jarrott both described the same incident where they, as motorists, were horsewhipped by a passing trap-driver. Here, they were travelling in convoy to Ripley, Surrey, Jarrott piloting a De Dion voiturette with two friends, while J. W. Stocks (1871–1933) was driving Edge on a quadricycle. In response to the assault, Edge gave chase and mounted the trap, ripped the driver's clothes and thrashed him with his own whip.[92] In another example, in the winter of 1900, Edge, fellow ABC member Alfred

---

[86] Jarrott, *Ten Years of Motors and Motor Racing*, p. 24.
[87] Edge, *My Motoring Reminiscences*, p. 8; G. H. Smith, *Selwyn Francis Edge: The Man and Some of the Things He Has Done* (privately pub., 1928), p. 6. Jarrott related an episode where, as early motorists, they were horsewhipped by a passing trap-driver who was dealt with 'in the fashion most approved of by Englishmen': Jarrott, *Ten Years of Motors and Motor Racing*, p. 243. For a context of casual violence in late-nineteenth-century Britain, see John E. Archer, '"Men behaving badly?": Masculinity and the uses of violence, 1850–1900', in Shani D'Cruze (ed.), *Everyday Violence in Britain, 1850–1950* (Harlow: Longman, 2000), pp. 41–54.
[88] Jarrott, *Ten Years of Motors and Motor Racing*, pp. 242–4.
[89] *Cycling*, 7 October 1893, p. 182. My thanks to Damien Kimberley and Peter Card for this reference.
[90] *Motoring Illustrated*, 22 March 1902, p. 52. See also Kurt Möser, 'The dark side of automobilism, 1900–30: Violence, war and the motor car', *Journal of Transport History*, 24:2 (2003), pp. 238–58, at p. 248, who describes the stretching of a wire across a road in Germany in 1913 which decapitated two occupants of a passing car.
[91] *The Autocar*, 23 September 1905, pp. 371–2.
[92] Edge, *My Motoring Reminiscences*, pp. 217–8; Jarrott, *Ten Years of Motors and Motor Racing*, p. 243. No date is given, but it would be about 1899.

Nixon and Nixon's sister went to Brighton in Edge's Panhard-Levassor. On their way back, leaving after dark, a 'yokel' threw ice at them and hit Edge on the head, who then gave chase and caught the assailant, forcing him to strip. Edge then drove on and left the clothes with other 'yokels' in the next village.[93] As part of Mrs Kennard's trip on an 8hp Napier to St Andrews in 1900 (so that her husband could play golf), she encountered an oncoming gig. 'As usual, the driver occupied the entire road, and displayed no intention of budging.' Kennard's car skidded, but no one was hurt. 'These good folk jog along, sometimes half asleep, quite often in the land of Nod, or else immersed in the perusal of book or newspaper – that is, when they are sober. God defend the poor autocarist who encounters them. He is at their mercy.'[94] But on a trip later that year, as she was coming through Leicester, a thirteen-year-old boy threw a lump of mud, hitting her and her husband in the face. Their mechanic gave chase and the boy was traced to a house where the parents were obliged to 'inflict fitting castigation'. Meanwhile, Mrs Kennard, left on the car, was surrounded by a 'jeering crowd' with 'mocking faces'.[95]

Accommodation

The first cyclists, and especially motorists, once on the 'open road', found themselves fair game as they sought accommodation or refreshment. As the railways had taken passing traffic off the roads in the nineteenth century, this had a somnambulant impact on staging inns and hotels. Hissey, on his horse-drawn exploration of rural Kent in the 1880s bemoaned this development.[96] In his cycling diary, G. H. Smith described how poor the hotels and inns had been for the early cycling clubs in terms of the welcome, catering and accommodation they offered.[97] The motoring press published letters and articles on the underwhelming experience of the hotel when on tour. *The Car Illustrated* in 1905 drew on an extract from the *Ladies' Pictorial*: until the advent of the motor car, women knew little of the country hotel and village inn, whereas now she knew too much. Describing bad food and bare rooms, there is 'not to be found a touch of refinement, or a thought for her comfort. Men, with their incomprehensible glee at "roughing it" do not know what the women they rush through rural England in their beloved motors occasionally have to endure'.[98] *The Motor* reported on rural hotels and received a letter from J. Fletcher, who had just been charged nineteen shillings for supper and bed and breakfast: 'Being a man of very moderate means and motoring for health and pleasure, I could not stand such charges'.[99] Even accommodation in the bigger towns was risible: the *Motor Car Journal* commented on hotel accommodation during the Thousand Mile Trial of 1900 and found 'Some hotels in even leading towns

---

[93] Edge, *My Motoring Reminiscences*, pp. 215–7.
[94] *The Autocar*, 27 October 1900, p. 1035.
[95] *The Autocar*, 29 December 1900, p. 1259.
[96] Hissey, *A Holiday on the Road*, chap. 15.
[97] Smith, *Some Notes about the Anerley B.C.*, p. 8.
[98] *The Car Illustrated*, 6 September 1905, p. 79.
[99] *The Motor*, 4 January 1910, p. 885.

were beneath contempt' (for some reason, Sheffield especially).[100] The rooms were not cheap, either; in Manchester, of the prices listed for the thirty-three hotels suggested, it could cost from 6s 6d for a room at the Grand (and 3s 6d for one's servant) down to three shillings for a room elsewhere.[101] The journalist Alexander Bell Filson Young (1876–1938) cautioned against the gastronomic experience: 'English inns are not attractive enough to tempt the wise motorist to lunch at them, unless, indeed, he has the passion for cold beef, Cheddar cheese, green wineglasses, and flies.'[102] There was little improvement over time. Jarrott described the extremely uncomfortable feather bed when he put up at a country inn with two friends in 1912.[103] G. Thornton Bridgewater wrote to the motoring press in 1914 to explain how the motorist was still being fleeced. There were two scales of charges, one for the pedestrian, one for the motorist, he reported. 'I guarantee to say if you arrive at an hotel in a car your food will cost you not only more but just about twice as much as if you arrive by the hotel bus.' The same principle had extended to shops, and some motorists had learnt to park beyond the shop in order to appear as pedestrians 'and ensure being reasonably charged.'[104] Even as late as 1919, 'Observer' wrote to *The Motor* to describe his miserable hotel service. As a passing motorist he was offered for lunch indifferent soup, cold lamb, one potato and half a pear, whereas the residents were being served a hot lunch. All this, and the hotel was displaying AA and RAC signs,[105] which was meant to be a method of ranking accommodation for the benefit of the touring cyclist or motorist, a service available to club members.

There were exceptional hostelries, where new visitors were welcomed, and these quickly became legendary amongst cyclists and motorists. The Anchor Inn at Ripley, Surrey, run by Mrs Harriet Dibble (d. 1887, after which her daughters took over), had been a favourite for cyclists since the 1870s, situated conveniently close to London and offering hospitality all year round.[106] The Hut Hotel at Wisley was featured in *The Car Illustrated* as one of the most favourite stopping places for motorists on the London-to-Portsmouth road.[107] For motorcyclists, Ye Olde Thatched House at Epping was identified by *Motorcycling* as a 'popular motorcycling resort'.[108] It was also clear that some travellers had a knack for finding good accommodation: Charles Manners wrote in to *The Autocar* to record his positive experiences with hotels. He could

---

[100] *Motor Car Journal*, 28 July 1900, p. 355.
[101] Elizabeth Bennett, *Thousand Mile Trial* (Heathfield: Elizabeth Bennett, 2000), p. 167.
[102] Alexander Bell Filson Young, *The Happy Motorist: An Introduction to the Use and Enjoyment of the Motor Car* (London: E. G. Richards, 1906), p. 209.
[103] Charles Jarrott, '"Half-time"', *The Strand*, 44 (1912), p. 310.
[104] *The Motor*, 21 April 1914, p. 563.
[105] *The Motor*, 6 May 1919, p. 442.
[106] For a description of Ripley and the Anchor, see Carlton Reid, '"The Mecca of all good cyclists"', chap. in *Roads Were Not Built for Cars: How Cyclists were the First to Push for Good Roads and became the Pioneers of Motoring* (Newcastle on Tyne: Front Page Creations, 2014), pp. 136–44; Nicholas Oddy, 'The Anchor Hotel, Ripley: An analysis of the cyclists' books, 1881–1895', *Cycle History*, 13 (2002), pp. 108–13. Other favourites were the Angel at Ditton, the Bear at Esher and the White Lion at Cobham: Arthur C. Armstrong, *Bouverie Street to Bowling Green Lane: Fifty-five Years of Specialized Publishing* (London: Hodder, 1946), p. 58.
[107] *The Car Illustrated*, 22 March 1905, p. 148; and photograph in Morrison and Minnis, *Carscapes*, p. 279.
[108] *Motorcycling*, 13 September 1910, p. 451.

find a 'clean looking inn or small wayside hotel' outside a large town, he explained, and generally got 'capital entertainment, with dinner or supper, bed, bath, breakfast, drinks, everything included', for six or seven shillings a head.[109] Cyclists and motorists puzzled over why any publican or hotelier failed to see the benefits of making their new patrons welcome. The automobile-promoting *Daily Mail* reported 'extortions pure and simple' levied by hotel proprietors for the stabling of a motor car. 'It is as well that hotel-keepers should remember that no class of the community is likely to benefit them more in the near future than automobilists themselves, and that those who treat the tourist unfairly will only find their names on the black list of the Automobile Club.[110] Certainly, some proprietors did take advantage of their motorized customers, charging extra to 'stable' a motor-car, but other proprietors, particularly in the earliest days, would genuinely have believed the motor car to be dangerous and liable to explode (and therefore not welcome). Furthermore, proprietors too were dismayed at the way they were stereotyped. There was an occasional glimpse from their point of view, for example in *The Car Illustrated* in 1914: 'You must remember that many motorists, especially amongst the small car owners, are very stingy. I suppose it arises from the fact that to many a car is really a luxury and a hobby, and the motorist is spending everything he has spare or can afford, on his car, and has little margin.' The evidence was that motorists grumbled, did not tip, and some even came in for lunch or tea and left without paying.[111]

Clubs

The club was an integral part of sports culture in the nineteenth century.[112] Sports such as football and cricket were codified in the nineteenth century, enabling teams to play each other by the same rules, and national leagues to be established.[113] As new sports evolved, it was natural for clubs to form, to enable inter-club and national competition, but also to ensure the middle-class clubmen could associate with like-minded people of their own class and manners. (Horse-)driving, then cycling, and ultimately, motoring were seen as sports, and therefore required the formal mechanism of the club, with its rules, officers, uniform and code of conduct. An introduction was usually necessary and a high ongoing annual subscription ensured the right class of member. It was this exclusivity that worked for clubmen when their sports were minority interests, but once cycling and motoring started to trickle down the social scale, the club for

---

[109] *The Autocar*, 9 April 1910, p. 491.
[110] *Daily Mail*, 3 April 1903. The 'Automobile Club' is the Automobile Club of Great Britain and Ireland, formed in 1897. In 1907 it was renamed the Royal Automobile Club (RAC).
[111] *The Car Illustrated*, 11 February 1914, p. 4.
[112] See Adrian Harvey, *The Beginnings of a Commercial Sporting Culture in Britain, 1793–1850* (Aldershot: Ashgate, 2004); Mike Huggins, 'Sport and the British upper classes c. 1500–2000: A Historiographic Overview', *Sport in History*, 28 (2008), pp. 364–88.
[113] See, for example, Richard Holt, *Sport and the British: A Modern History* (Oxford: Clarendon Press, 1989).

them became increasingly irrelevant. The values of the clubs tell us much about social stratification in the late-nineteenth century.

The Four in Hand (horse-)driving club had been formed in 1856 and, as was conventional, imposed a uniform, rules and a social calendar with its formal dinners. This was the same with the Road Club and the Coaching Club, formed in 1870.[114] The coaching revival followed a similar model. While it was run by proprietors and professionals, amateurs also paid a subscription to drive the coaches.

Cycling clubs followed much the same pattern. The Catford Cycling Club, formed in 1886, for example, had an etiquette for its club runs, with a strictly enforced uniform with distinguishing marks for the captain and vice-captain. The captain was to cycle at the front and could not be overtaken. Riders would be expected to learn and abide by a set of signals transmitted by bugle or whistle, and this called for a nominated bugler; Preston Cycling Club once briefly endured the ignominy of not having any member able to play the bugle.[115] Indeed, the use of the bugle as a means of warning of imminent arrival was entrenched, having been adopted by horse-driving clubs too. As a boy, the chauffeur Martin Harper (b. 1880) could remember how the postman would alert his father, the sub-postmaster, of his approach by blowing 'a little tune' on his bugle.[116] When motoring pioneer Claude Johnson (1864–1926) arrived in his motor car at his hotel in the early hours of Christmas Day 1899, the landlord, having been knocked up, sounded his bugle to inform the whole community of the event. Also, Edge, once he took to breeding pedigree pigs in the 1910s, summoned them by bugle for feeding.

The ABC was probably typical in the formality of its club tours, say the regular weekend trip, which required captains, buglers, uniforms and discipline.[117] The club also enjoyed lavish dinners with its associated rituals, requiring full dinner dress. Patrons, presidents and senior members were placed on the top table, meaning that a strict code of status within the club was to be observed; Edge, for example, could expect this once he became patron or president of numerous cycling and motoring clubs. The Motor-Cycle Club dinner in 1904 was held at Frascati's, a fashionable restaurant noted for its cuisine, on Oxford St, London (and to which the ABC also moved in 1929). With Edge as president (he was one of the founder members in 1901), he was accompanied by other cycling–motoring celebrities such as Mark Mayhew (1871–1944) and Stocks;[118] a photograph in *The Motor* showed a well-mannered party in dinner dress and seated at their tables.

The financial demands made on members of the earliest cycling clubs also meant membership was self-selecting. The Rossendale Bicycle and Tricycle Club, founded in 1878, for example, charged an annual subscription of around 7s 6d, plus a uniform which was compulsory. The Blackburn Cycling Club uniform in the early 1880s was

[114] Beaufort, *Driving*, pp. 260–6.
[115] Zoe Lawson, 'Wheels within wheels: The Lancashire cycling clubs of the 1880s and 1890s', in A. G. Crosby (ed.), *Lancashire Local Studies in Honour of Diana Winterbotham* (Lancaster: Carnegie Publishing, 1993), pp. 123–45 at pp. 124–5.
[116] Martin Harper, *Mr Lionel: An Edwardian Episode* (London: Cassell, 1970), p. 3.
[117] Smith, *Some Notes about the Anerley B.C.*, p. 49.
[118] *The Motor*, 13 January 1904, p. 595.

Figure 2.1 A cycle club out on tour, c. 1880, with a mixture of 'ordinaries' and tricycles. This predates the arrival of the 'safety' bicycle. Illustration by Joseph Pennell, in Viscount Bury and G. Lacy Hillier, *Cycling* (London: Longmans, Green, & Co, 3rd ed., 1891), opp. p. 194.

thirty-two shillings, while the grey stockings of the Vale of Lune Cycling Club were 7s 6d.[119] Malvern Cycling Club charged an entrance fee of five shillings and an annual subscription of the same, while its uniform of coat, knickerbockers and stockings cost around fifty shillings.[120] In true amateur style, a culture of prize-giving (rather than cash prizes) flourished. An amateur might pay a fee, or settle expenses, to enter a race, but would not expect a cash reward. Plus, it was usually only the sportsman with time on his hands who could enter for long-distance endurance records.

There was social credit to be gained by being associated with a club, just as the number of clubs that one joined was also indicative of one's social status: Edge, for example, was a member of several cycling clubs. It was a sign of prestige for a club when a 'crack' chose to wear the colours of a particular club; at the end of 1889, it was announced that Edge, by then a cycling record-breaker, intended to ride in the Catford Cycling Club colours for the 1890 season – for the club, this was 'welcome news to the majority of members'.[121] Club members were usually 'something in the city',[122] and this in turn helped attract highly connected patrons; Catford attracted patronage by lords, MPs, clerics and JPs. Patrons, in turn, would be expected to endow the club with cups

[119] Lawson, 'Wheels within wheels', pp. 124–5.
[120] Roger Alma, *The Malvern Cycling Club, 1883–1912: A History of Cycling in Malvern, Worcestershire, before the First World War* (John Pinkerton Memorial Publishing Fund, 2011), pp. 17, 20.
[121] *Licensed Victuallers Mirror*, 3 December 1889.
[122] Southcott, *The First Fifty Years of the Catford Cycling Club*, pp. vii–ix, xi.

and awards, to be presented at each annual dinner; here, the ABC benefited from the largesse of Edge as its president in the 1920s and 1930s, hosting the 'Edge's run' and allowing club members to stop off for tea at his country home.

Club membership brought the responsibility of appropriate behaviour when representing the club: fair play, observance of social etiquette, observance of rules and codes of conduct. It was understood that one's club should be publicly respected. Cycling in club uniform and, initially, motoring as a club, were unacceptable on a Sunday; Alfred Cornell remembered how Sundays were strictly observed 'to give a proper impression that those interested in motor-cars were devout and righteous men'.[123] Walter Groves (b. c. 1866), later on the editorial team of *Cycling* and *The Motor*, recalled how his early cycling was handicapped by his neighbours' objecting to seeing him out cycling on a Sunday, so he left his bicycle at a friend's house on a Saturday night ready for the morning. Indeed, in the early days of *Cycling* (launched 1891), 'it was not permissible to refer to any happening or club run as having taken place on the seventh day of the week'.[124] Smith recalled the worry, in 1891, over whether twelve-hour cycling races should be conducted on a Sunday.[125] Clubs saw their activity as a 'sport', and often did their own organizing, setting and authentication of record runs,[126] a feat of organization gleaned, for some, through the military training at school. To do this required an army of volunteers, as time-keepers, look-outs or pacers. The clubs had to present an efficient public face, as the cycling and motoring press reported the organization of club events, and this then reflected directly on the standing of the club nationwide.

For those able to secure an entrance, the rewards of club membership were terrific. The clubs did not forget their own: the cyclist-turned-motoring entrepreneur Frank Shorland (1871–1929), for example, was honoured in 1924 by his old club, the North Road Cycling Club, in celebration of the thirtieth anniversary of the 'completion of his career as a racing cyclist'; this involved a dinner at the Connaught Club with toasts, speeches, and a programme of music.[127] Edge was another who, as a member and, later, patron of cycling and motoring clubs, gained life-long friends and future business associates, and access to highly connected individuals, information and technical skills. He enjoyed the camaraderie that came with the regular runs, the trials and races and the holidays, and a sense of belonging and mutual protection. While the cycling and motoring clubs facilitated sport, and camaraderie and protection in numbers, one club was established specifically as a commercial talking-shop for entrepreneurs in both the cycling and the motoring worlds: the Wheel Club was formed in 1910 to provide a meeting-place for the manufacturers and agents in the cycle trade. Securing rooms at the Inns of Court Hotel in London, it charged one guinea for town members

---

[123] Cornell in [Lord] Montagu, *The First Ten Years of Automobilism, 1896–1906* (London: *The Car Illustrated*, 1906), p. 45.

[124] Armstrong, *Bouverie Street to Bowling Green Lane*, pp. 2, 8.

[125] Smith, *Some Notes about the Anerley B.C.*, p. 35. It was 1924 before the Club changed to race on Sundays, long after other clubs had made the switch.

[126] See Alfred Nixon's efforts to authenticate his Land's End to John O'Groat's record attempt in 1885 by identifying to the cycling press the gentlemen who would support him en route: *Bicycling News*, 2 September 1885, p. 555.

[127] 'Complimentary dinner to Frank W. Shorland', 30 September 1924: University of Warwick, Modern Records Centre, MSS.328/N16/4/1.

and half a guinea for country members, with an entrance fee of one pound. For this fee, members had access to a spacious clubroom, a card room and a drawing room. The *Cycle and Motor Trades Review* believed that membership would be 'an excellent investment merely on account of the money which members would save by the club tariff alone', recognizing the value of association with like-minded business contacts. Club socials also presented opportunities for businessmen to make useful contacts in their trade. Entertainment included, to take the example of the evening of 2 March 1910, an exhibition billiards match between amateur champions Lewis Stroud and Harry Davis, then a sparring (boxing) match between cycling-record holders 'Sammy' Bartleet and B. H. Hogan, with cyclist Harry Green as referee. ('Honours' appear to have been even.) With 'no less a person than' Edge in the chair, the soprano Miss Ada Tunks finished with some 'excellently rendered songs', evoking 'great applause'.[128] Having Edge present was good for his profile and put him amongst his own.

Class in cycling and motoring

The wider point remains, though, that the need to be in a club served cycling and motoring only as minority sporting pursuits, as they both were in their earliest days. Once a wider social group sought membership, as happened before the First World War, the club appeared increasingly irrelevant. The club, then, presented a barrier for widening participation in cycling and, particularly, motoring. In the years to 1914 a changing, and socially broadening, motoring demographic was slowly leading to fewer motorists seeking club membership. The Automobile Club recognized the incongruity of its mixed membership in its early days. Its historian Brendon has written:

> The aristocracy of automobilism [the Automobile Club] felt ambivalent about middle-class motoring. [. . .] It was proud that its pioneering labours had borne such copious fruit. But it was concerned that what had been an exclusive pleasure should now be pursued by the bourgeois multitude, who adopted the 'week-end' habit, preferred golf to church and aped the fashions of their betters.[129]

When the Club took drastic measures to bolster membership during the First World War, by halving the entrance fee and annual subscription, Edge wrote to *The Autocar* to point out that the Club would be overcrowded with newcomers unable to afford the dining room. 'Is this the sort of man for whom the Club, especially with the present clubhouse, was intended?' he asked. Instead, the Club '*must* be a commercial undertaking'. Members were uneasy about 'an immense influx of new blood', but aghast that it should become what Edge suggested, 'a creature of the trade'.[130]

---

[128] *The Cycle & Motor Trades' Review*, 10 March 1910. See also the photograph of the boxing match, caption: 'B.H. Hogan v. "Sammy" Bartleet, Referee, Harry Green, at The Wheel Club, 2nd March 1910': E. A. Penn, 'The wheel club', *The Boneshaker*, 114 (1987), pp. 4–5.
[129] Brendon, *The Motoring Century*, p. 10.
[130] Brendon, *The Motoring Century*, pp. 184, 186.

The Automobile Club had been founded to promote the new motor traction 'by arranging trials, fostering touring, teaching driving, issuing road maps, dispensing propaganda, approving garages and hotels, organizing insurance and legal assistance' and, eventually, road patrols. From 1899, following a struggle with the National Cyclists' Union it had become the governing body of motor sport. As a gentleman's club, though, an issue with the nature of its membership immediately arose. Edge had written to the Automobile Club's *Journal* in 1899 urging that anyone should be able to join 'providing there is nothing to be said against their character'. But club membership had been, and remained, highly selective. It was also very male: ladies were not allowed as members, and as guests could only enter 4 Whitehall Court (the Club's premises until 1902) between 3 o'clock and 6 o'clock in the afternoon (and then they were confined to the General Reception Room). And of the men, the Club's first full-time secretary Claude Johnson had said that '[a]nyone interested in automobilism is eligible for membership, though, naturally enough, we observe some social restrictions'. These included 'working managers' and 'professional cycle riders' (or, 'maker's amateurs'), which, in fact, neatly described Edge, but whose membership was a given from the outset: in the 1890s he was simultaneously racing for a cycle company while being employed as a travelling sales representative. (Ironically, Edge's brother Kelburne [1869-1925] was expelled from the Sydney Bicycle Club in 1893 for taking part in a cash race.)[131]

There was, then, some elasticity in the Club's restrictions: some of its earliest members, of modest background, included journalists such as Harmsworth (who bankrolled the Club's Thousand Mile Trial) and businessmen such as Harvey du Cros (1846-1918, who had an array of commercial interests); Edward Shrapnell-Smith (1875-1952), the one-time manager of the Road Carrying Company; and the Club's benefactor Paris Singer (1867-1932), who had founded the City and Suburban Electric Carriage Company. To complicate the mix, some of its early members, noble by birth, exploited their social situation to sell cars, notably Charles Rolls (1877-1910) and Montague Grahame-White (1877-1961), while Edge admitted in later years that 'he had made very good use of the Club's facilities for demonstrating the merits of the goods I had to sell'. With such a disparate membership, the Club was a hotbed of tension: Lord Montagu of Beaulieu (1866-1929) was in dispute with it because his society motoring magazine *The Car Illustrated* competed with the Club's *Journal*, while a simmering resentment also persisted amongst some of the provincial clubs whose members had to pay a half-guinea affiliation fee.[132] Edge had already foreseen, as early as 1899, that for the Automobile Club '[t]here must be quite a large and growing body of Automobilists who will be outside our ranks and who will then sooner or later

---

[131] Brendon, *The Motoring Century*, pp. 39, 40, 42, 44, 46. For Kelburne: *The Referee* (Sydney, NSW), 8 February 1893. Kelburne's club career was not over, though: he was made an honorary life member of the [Launceston] City Cycling Club in 1895 for supplying the medal for the 10-mile record: https://trove.nla.gov.au/newspaper/article/233093981. My thanks to Rosemary Sharples for these references. Reference accessed 17 July 2019.

[132] Brendon, *The Motoring Century*, pp. 10, 39, 40, 44, 48, 58, 60, 68 [Edge, reported from *The Autocar*, 13 November 1915, p. 612], 88. Brendon's is the finest of several histories of the national motor-car clubs; others for the Automobile Club/RAC include Raymond Flower and Michael Wynn Jones, *One Hundred Years of Motoring: An RAC Social History of the Car* (London: RAC Publishing, 1981).

Figure 2.2 The eliminating trials for the 1903 Gordon Bennett Cup were held in the grounds of Welbeck Abbey, Notts. The picnicking drivers in the foreground are (left to right) J.W. Stocks, Roger Fuller, Mark Mayhew, Charles Jarrott, S. F. Edge, W. T. Clifford Earp. Image: David Hales.

wish to found an opposition organisation'. This was addressed shortly afterwards, in 1901, when the Motor Union was formed as a spin-off by the Automobile Club, as an organization which any motorist could join, and motor magazines recommended that their readers should join it. The 'problem' of speed-trapping persisted, though, so to counteract this, the Automobile Association (AA) was formed in 1905, initially as the Automobile Mutual Association. This was in part following an initiative by motor agents Charles Jarrott and William Letts (1873–1957), whose company imported De Dietrich and Oldsmobile motor cars amongst others, and who had employed 'scouts' on bicycles to alert motorists on the Brighton Road of speed traps ahead. The AA, considered further in Chapter 7, adopted a similar strategy to the Motor Union and similarly attracted a broad social base.[133]

Another initiative was the formation of the Society of Motor Manufacturers and Traders (SMMT) in 1902. This was intended to divert the trade elements away from the Automobile Club, but 'did nothing to stop dissension between amateurs and professionals in the Club'.[134] Here, Simms was an important facilitator as he had been five years earlier with the formation of the Automobile Club. The entrepreneurs

[133] See Brendon, *The Motoring Century*, chaps. 3, 4. For the AA, see Ernest Stenson-Cooke, *This Motoring: Being the Romantic Story of the Automobile Association* (London: Automobile Association, 1931); David Keir and Bryan Morgan (eds), *Golden Milestone: 50 Years of the AA* (London: Automobile Association, 1955); Hugh Barty-King, *The AA: A History of the Automobile Association 1905–1980* (Basingstoke: Automobile Association, 1980).
[134] Brendon, *The Motoring Century*, pp. 88.

and industrialists amongst its original subscribers were Simms, the engineer Henry Burford (1867-1943), the motoring agents Letts, Jarrott and Edge, the engineer Sidney Critchley (1866-1944) and racing driver Henri Farman (1874-1958).[135] Most of these were drawn from backgrounds in cycling competition or sales.

The playing out of class distinctions within the membership of the Automobile Club was fully evident within cycling and motor traction. It also served as a barrier to dissuade potential motorists, thinking they would not, or could not afford to, fit in. Class as a means of social categorization in the late Victorian period still holds; Steinbach, for example, has defended the importance of class to late Victorians, describing their society as 'fiercely hierarchical' with class being 'simultaneously economic, cultural and discursive'. Here, occupation might play a more important part in one's status than one's income.[136] Social categorization and discrimination was no more evident than in that early cycling and motor world. 'The Pilgrim', a regular columnist in the magazine *Motorcycling*, dedicated a column in 1910 to 'The "swagger" motorcyclist'. 'Although grubby people on home-assembled mounts are often to be seen on the roads', he explained, 'motorcycling is at yet far from that comfortable five-shillings-a-day tour-as-you-like stage long reached by its sister pastime [cycling].' Therefore, 'many with slender purses are kept from becoming motorcycling tourists, and many others, already motorcyclists, do not tour because they are not in a position to do it in what appears to be considered "style", their resources being perhaps already pretty well taxed, if not actually strained, in purchasing and running their machines'.

That said, 'The Pilgrim' continued, it was undeniable that motor-cycling was still practised with 'a great deal more swagger and side' than pedal-cycling. 'The would-be entrant is impressed by [. . .] the rider's get-up and accessories' and knows what it would cost to emulate this. He concludes that to take up motor-cycling 'would only exhibit the limited nature of his means, and being unwilling to brave covert social sneers, he remains outside our ranks, where he can keep his pride and pretend he has no fancy for the pastime'. And similarly, for someone who had bought in to the movement, 'a similar impression of the social obligation to motor in style or not at all keeps him from touring much'.[137]

Indeed, in the idealized drawings on the covers of and within motor-cycling magazines, the motor-cyclist was always wearing the best gear. Expensive clothing for the motor-cyclist and motorist could be bought by mail-order from up-market shops such as Gamages, which advertised frequently, often over a full page, in those same magazines. For ladies, the 'Desiree' mica mask veil and hood 'as worn by the Countess of Dudley, and all society lady motorists' was available for twenty-one shillings; while a chauffeur's black leather jacket and trousers could be had from 46s to 95s. (The cap was extra.)[138] Dunhill's too, was famous for its 'Motorities', including the 'Swift hood'

---

[135] Nixon, *The Story of the S.M.M.T.*, p. 27.
[136] Susie L. Steinbach, *Understanding the Victorians: Politics, Culture and Society in Nineteenth-Century Britain* (London: Routledge, 2nd ed., 2017), esp. pp. 123-43. See also Andrew Miles, 'Social structure, 1900-1939', in Chris Wrigley (ed.), *A Companion to Early Twentieth-Century Britain* (London: Blackwell, 2003), pp. 337-52.
[137] 'The Pilgrim', 'The "swagger" motorcyclist', *Motorcycling*, 30 August 1910, pp. 387-8.
[138] *The Motor*, 8 July 1903, ads: Gamage's full page advertisement for clothing.

in 'Macintosh material' for twenty-one shillings (the gauze curtain 3s 6d extra) and 'Napier goggles' (3s 6d).[139]

Being able to afford the new motoring and motor-cycling clothing was one thing, but the spectacle of parading it also risked ridicule. *Punch* magazine, for example, lampooned the attempts by motorists, male, but especially female, to wear clothing that was both protective (against the dust, cold and rain) and dignified and fashionable. Even the motoring press recognized that there was a credibility problem here. *Motorcycling and Motoring* said in 1902 that 'many motorcars are now being fitted with a dust and wind screen, which consists of a celluloid window fixed upright on the front of the car. With this fitted, there seems no necessity for the occupants of the car to wear hideous face masks'.[140] (Windscreens on motor cars were usually available as expensive and optional extras.) The more expensive motoring magazines such as *The Car Illustrated* had regular columns on ladies' motoring fashions, endorsing some products, while Gamage's, with their 'Sportswoman' hat, claimed to have 'solved the problem of having useful headwear without the usual accompanying hideousness'.[141] In 1902 Lady Mary Jeune (1845–1941) tackled 'the question which affects women very deeply', namely, how to dress to motor in warmth and comfort 'with as little disfigurement as possible'.[142] *Punch* also reflected a wide general unease with the adoption of 'rational' dress by women in their pursuit of greater comfort when engaging in sports.[143] 'Swiftsure', the cycling correspondent for the *Clarion*, recorded in 1895 how 'few would believe how insulting and coarse the British public could be unless they had ridden through a populated district with a lady dressed in Rationals', while ladies were encouraged to ride bicycles in an upright, 'ladylike' manner (that is, not bent over the handlebars showing any exertion).[144]

Even having all the right clothes and being able to afford the finest motor cars was not necessarily enough to pass muster in such a socially sensitive environment. Edge took great care over his public appearance, and his overt displays of wealth and success were based on real achievement and toil. Yet, a drawing by 'The Tout' in *Tatler* magazine in 1914 revealed much about his standing amongst his peers. The drawing showed eighteen men, all 'distinguished motorists', mingling at Brooklands racetrack. They included peers, MPs, fellow entrepreneurs and racing drivers such as Harvey du Cros, Charles Jarrott, Frank Shorland and Edge. Each man was caricatured, as was the fashion in magazines at the time. With the exception of Gerald Biss (1876–1922), author, 'motoring expert and man of many friends',[145] and seen in the picture airborne in a kite, Edge is the only one depicted as odd. He is represented as a spiv, and, despite

---

[139] *The Autocar*, supplement, 11 November 1905.
[140] *Motorcycling and Motoring*, 6 August 1902, p. 425.
[141] *The Car Illustrated*, 7 December 1904, p. xiii.
[142] [Lady] Jeune, 'Dress for motoring. I. Dress for ladies', in Alfred Harmsworth (ed.), *Motors and Motor Driving* (London: Longmans, Green & Co, 1902), pp. 66–71.
[143] For recent literature on rational dress, see Jungnickel, *Bikes and Bloomers*. See also Tracy Collins, 'Athletic fashion, "Punch", and the creation of the New Woman', *Victorian Periodical Review*, 43 (2010), pp. 309–35; 'What it will soon come to', *Punch*, 24 February 1894, p. 90.
[144] Lawson, 'Wheels within wheels', pp. 123–45 at pp. 134–5. See also Erskine, *Lady Cycling*.
[145] *The Times*, 17 April 1922.

his height – well above average – is seen as short and shifty.[146] Here he was truly the wheeler-dealer, moving around his contacts and prospects. Furthermore, a few years later, his record for the war effort turned out less than glorious, probably as a result of his failing to recognize when to be diplomatic with his superiors. Many of his peers had helped in the war effort: Jarrott (inspector of transport for the Royal Flying Corps) came out with an OBE, Percival Perry (1878–1956, chief assistant to Sir Arthur Lee, the Director-General of the Food Production Department) was knighted. Yet, Edge, as Director of Agricultural Machinery reporting to the Board of Agriculture, was sacked.

'The Pilgrim', writing in a subsequent column in *Motorcycling*, commented on the otherwise unspoken 'pecking order' evident on the road. Describing the relations of motorcyclists with cyclists and motor-car drivers, 'Many of us are apt to treat the pedal-cyclist very much like a poor relation', and they return 'our contemptuous or lordly glances with scowls, with assumed indifference, and if we appear to be *en panne*, with undisguised gloating'. On the other hand, we regard the man 'who can loll in the comfort of a car' 'with deference'. We [motorcyclists] have come to be called 'motorists', treated as 'motorists', supposed to have the same tastes and wants. We are looked after by motoring bodies 'who do not understand our needs and do not think it worthwhile trying'.[147] Meanwhile, reports of aggressive behaviour were common. 'A member of the Motor Cycling Club' described the moment, when, on the club run to Brighton in 1905, he was driven at by a car driver and had to skid off the road. He also described a similar event with a 40hp Mercedes being driven on the wrong side of the road on a bend.[148] A common representation of motoring involved the fear it created, particularly of being run over,[149] but fear on the road could come from all quarters: if it wasn't motorists, then cyclists would 'scorch' and menace, as the 1910 *Punch* cartoon 'Terrors of the road' suggested.[150]

Conclusions

Harry Lawson had his own interests in mind when, in 1896, he made his bold predictions about the demise of the horse in the face of the new motor traction. He was ultimately proved right, but there were many hurdles to leap first. These include Lawson himself, whose attempt to control the patents pertaining to motor traction was too aggressive for the entrepreneurs, customers and the fledgling motor industry. Furthermore, his association with other crooked financiers in huge company flotations relating to the bicycle and motor industries served as an additional brake.

Lawson, though, was giving his speech on the very day that the restrictive 'red flag' legislation was repealed, and it is curious that the removal of this impediment had little

---

[146] Image from *The Tatler*, 29 April 1914, in 'Laurie Cade looks back, part 2', *Veteran and Vintage*, 5:5 (1961), pp. 151, 166.
[147] 'The Pilgrim', 'The Relations of Motorcyclists with Pedal cyclists and Motorcar Drivers', *Motorcycling*, 13 September 1910, p. 451
[148] *The Motor*, 11 April 1905, p. 277.
[149] Imagery in Jeremiah, *Representations of British Motoring*.
[150] 'Terrors of the road', *The Motor*, 27 December 1910.

immediate effect. Most vehicles on the British roads were foreign at that point, and that remained the case. Entrepreneurs worked on encouraging a domestic production – most notably Edge and his promotion of the Napier brand – but they also continued to import vehicles, no matter how much commentators at the time might have appealed to a patriotic spirit. After all, the British had had a firm grip on cycle manufacturing but were unable to convert this to advantage with motors.

The existing system of horse traction offered certainties – of arrival, for one – which the motor vehicle clearly could not. Cycling and motoring were appearing as additional sports for the middle classes to indulge, rather than a system of mobility to supersede the horse – that was to happen slowly in the years up to and beyond the First World War. Horse-driving persisted as a sport – indeed, the Ranelagh gymkhanas up to the turn of the twentieth century continued to stage exhibitions of horsemanship on the coach-and-four. The coachbuilder was reluctant to move to motor traction and the carriage remained the pinnacle of status. There was no particular enthusiasm for those with equestrian skills or trades to retrain, in part fed by an aversion to technology and change.

It was also becoming clear to cyclists, and then motorists, that theirs were sports which potentially placed them in a situation where they were uniquely vulnerable. The roads were adequate for horse traction, but hazardous for cyclists and motorists, particularly with the speeds at which they could potentially travel. As with most sports, cyclists and motorists needed to pool together, the usual expression being the club. This meant a set of codes and rules which could not allow for any incursion by a wider group, that is, one that was largely unclubbable; cycling and motoring as sports could not then translate directly into a wider mobility option. The pioneer motorist, as the pioneer cyclist before, had taken his sport, and its imposition on others, for granted, or with a sense of entitlement. The reaction was often one of violence and aggression by a wider population of road users who saw themselves as increasingly disenfranchised. Meanwhile, accommodation on the move remained poor, with proprietors, to the dismay of cyclists and motorists, reluctant to embrace an additional line of business.

The late-Victorian period was entrenched in a class system which was apparent from one's clothing, accent or job, and this translated directly into the motor-traction movement. Cyclists, for example, might decline to 'move up' to motor-cycling if they felt they were being graded by what special or fashionable clothing they could afford. The same cyclist might have seen little advantage in seeking to join a club with its rules and Sabbatarian values. Potential motorists continued to feel alienated in an environment that threatened to expose them as parvenus; this might be by being seen as not able to afford the right clothing, or not being able to tour overnight because of the hotel rates. Letters in the motoring press recorded just how expensive a day out could be, with the hotel catering and tips. A similar hierarchical system existed within the movement, with cyclists feeling they were being sneered at by motorcyclists, while they and light-car owners were at the mercy of the aggressively driven large car. While clubs in general adjusted their behaviour – dropping the requirement for a uniform, allowing women as members, and so on – the social stigma remained. Any attempt to present motor traction as acceptable and respectable – and accessible – was to be welcomed, and the entrepreneur played a key role hereand which forms the theme of the next chapter.

3

Entrepreneurs

The novelist Mary Eliza Kennard (c. 1844–1936) specialized in a genre of 'sporting' fiction that was very much in vogue at the end of the nineteenth century.[1] As 'a keen cyclist', 'an adorer of tandem riding', a pioneer motorist and sportswoman,[2] it was natural for her to latch on to the sport of motoring as a fashionable theme for her novel *The Motor Maniac* of 1902. This was a semi-autobiographical account of a Mrs Janet Jenks's determination to buy and learn to drive a motor car. The story serves as a remarkable, if contrived, snapshot of the culture of the motoring 'movement' in its early days, placing us in what is seen now as a world defined by gender boundaries and almost suffocating codes of class. Her middle-class readers, having paid six shillings for the book, knew what they were getting: a tale of glamour, sport, cheating, and beating the foreigners through pluck and hard graft, all couched in the exciting new world of the motor car.[3]

Kennard also described exchanges with dubious motor-car salesmen, and she was not the only fiction writer inspired by personal experience. In her autobiography, the American novelist Alice Williamson (1858–1933) described Harvey, a cad whose 'ancestry was dotted with Dukes; he looked as if he had stepped from a portrait by Velasquez and changed the costume of that time at a smart tailor's. He was so grand that he considered most kings nouveaux riches, and hardly gentlemen.' Williamson traded in her crashed Benz, with Harvey selling her a motor car which 'really look[ed] as handsome, as debonair as Harvey himself', but soon 'showed its true nature, for what it was: a beast, a brute, a curse'.[4] The experience fed into her first novel, *The lightning conductor: The strange adventures of a motor-car* (1903), co-written with her husband Charles (1859–1920). Here, the young American heroine is dazzled by Reginald Cecil-Lanstown, who sells her a dud motor car and a dreadful chauffeur to go with it. The book had glowing reviews (mostly by motoring magazines),[5] sold all over the world

---

[1] For a discussion of motoring literature see Mom, *Atlantic Automobilism*, chap. 2, esp. pp. 170–6 (Williamsons); and p. 145 (Kennard).
[2] *Bicycling News and Motor Car Chronicle*, 2 May 1900, p. 19.
[3] Mrs Edward Kennard, *The Motor Maniac* (London: Hutchinson and Co, 1902). Esme Coulbert also discusses Kennard and her fiction in 'Perspectives on the road', chap. 5. A review of Kennard's *The Motor Maniac* concluded that 'readers who do not possess an automobile will be merely exasperated by the book' (*The Spectator*, 24 January 1903, p. 23).
[4] Williamson, *The Inky Way*, pp. 59–62.
[5] 'An ideal romance of the car' which 'expresses admirably the fascination of the open road', pronounced *The Car Illustrated*. It 'can be read with pleasure by motorist and non-motorist alike': *The Car Illustrated*, 3 December 1902, p. 44.

(a million in the United States), and sealed the Williamsons' reputations as writers of motoring fiction.

The journalist Alexander Bell Filson Young (1876–1938), then writing in 1906, similarly deplored the early motor trade:

> I do not know any trade that in the competition and struggle of its early days fell so low as the motor trade. I mean morally low [. . .]. Any impartial person taking up a trade motor paper a few years ago must have been shocked by what I can only describe as the frantic lying of the advertisers: assertions and counter-assertions, boasts, denials, hints, back-handed hits at rivals – all the dreary dust of the advertising system thickened into a very fog of darkness and untruth. The bewildered buyer in those days turned from this parade of infinite promise and guarantee to find, once he was fairly in the hands of the agent or manufacturer, that he had merely stepped from the frying-pan into a roaring furnace. To untruth was now added carelessness, incapacity, and in too many cases actual and flagrant dishonesty; he was overcharged, cheated, robbed right and left.[6]

The motor-car salesman, then, has never had a good reputation. Those involved in the earliest days of motor-traction sales were themselves often entrepreneurial, from a world of business, and who in turn had never been popular; Ashton has described the stereotype in the nineteenth-century literature as 'ostentatiously rich or poor-spirited and mean'.[7] Yet their reach and influence were increasingly important. Porter, for example, has shown the interconnection between sport, remuneration and the entrepreneur.[8] McGoun has recognized that the entrepreneur of the new motor traction had the most to gain by encouraging racing and ensuring that the glamour followed it. For publishers of motoring magazines, any motoring activity (racing, shows, 'gymkhanas', club days out) provided copy, which in turn created more interest and demand from readers. Agents such as S. F. Edge ensured journalists were given rides in cars on the trials, as this led to a story in the press, often with photographs. Organizing these activities was costly and would be undertaken only if it meant the promoters who backed them would see a return; this could be the creation of a brand in the minds of the public, or the association of a 'make' with qualities such as reliability or speed. This suited Edge, then, by seeing a return by way of sales.[9]

---

[6] Filson Young, *The Happy Motorist*, pp. 28–30. He was talking of only a few years before.
[7] T. S. Ashton, 'Business history', in Richard Davenport-Hines (ed.), *Capital, Entrepreneurs and Profits* (London: Frank Cass, 1990), pp. 8–9. See also Harrison, 'Growth, entrepreneurship and capital formation'.
[8] A good starting point is D. Porter, 'Entrepreneurship', in S. W. Pope and J. Nauright (eds), *Routledge Companion to Sports History* (Abingdon: Routledge, 2010), pp. 197–215. See also Ritchie, *Early Bicycles and the Quest for Speed*.
[9] I am grateful to Prof McGoun for making available to me his unpublished paper, 'Automobile commerce and competition in the nineteenth century', delivered at the 4th Annual Michael Argetsinger Symposium, Watkins Glen, 9 November 2018. See also Merki, for whom the persistence of motor traction was all down to racing and its social capital: Christoph Maria Merki, 'The birth of motoring out of sport: Car racing as a public relations strategy, 1894–1905', in Laurent Tissot and Béatrice Veyrassat (eds), *Technological Trajectories, Markets, Institutions: Industrialized Countries, 19th–20th Centuries* (Bern: Peter Lang, 2001), pp. 227–49.

Figure 3.1 *Punch* magazine initially lampooned the motor car as uncontrollable, and the motorist as hapless. This cartoon is by Leonard Raven-Hill: *Punch*, 23 January 1901, p. 63. Image: Richard Roberts Archive.

Filson Young may have had in mind cold-calling salesmen such as Percy Richardson (*c.* 1878–1955), then of the Great Horseless Carriage Co. In 1897, given a Daimler to sell, Richardson established that most of the wealthy prospects were shooting in Scotland, so, without any enquiries or introductions he drove the Daimler from Coventry to Edinburgh, a journey that took over a week. Identifying likely candidates from *The Scotsman*, he went to the estate of 'Sir Charles —', and (literally) put his foot in the door when the butler attempted to shut it on him. On this occasion it was worth it, as he ended up giving all the house party trial runs, was asked to stay on a week, sold the car and took deposits for six more, and in four months had orders for £27,000.[10]

The methods of Edge were a little more subtle: 'My customers would mention their cars to their friends', he said. 'I would get invitations to country houses, sometimes as far from London as Lord Lonsdale's seat, Lowther Castle [Cumbria], being asked to motor up, to shoot and stay; but actually, as I knew, the idea was that this S. F. Edge would come in one of his motorcars, and the bare fact that I should do this made me a welcome guest.'[11] As with Richardson, though, Edge was clearly the interloper, doing

---

[10] R. D. F. Paul, 'Selling cars in the late '90s: The trials of a pioneer salesman thirty years ago', *The Motor*, 29 January 1929, pp. 1246–9. The article also identified Edge, Walter Bersey (motoring designer), Sir William L. Sleigh (one-time president of the Scottish Motor Trade Association), Montague Grahame-White, William Letts (motoring entrepreneur), Lawson, J. S. Critchley (motoring works manager), Percival Perry (motoring entrepreneur), Ernest Instone (motoring works manager) and J. W. Stocks (racing cyclist-turned-motoring entrepreneur). See also E. K. S. Rae, 'Unconventional portrait of the week: Mr F. W. Shorland', *Automotor Journal*, 9 April 1910, pp. 383–5.

[11] S. F. Edge, 'Looking back: Interesting comparisons of yesterday and today by a pioneer', *Morris Overseas Mail*, May 1931, pp. 7–9.

the unsavoury task of selling. 'In those days, up to say – 1900, nobody but a well-to-do man could afford a motorcar,' he said. 'People who had them to sell did not, in many cases, believe in them. Therefore they said: "Let us put on a good profit, because we shall never sell *this* man another!"'[12]

Edge was from a commercial middle-class family, and apart from some private tuition, his education was undertaken at the modest Belvedere House College, at the end of the same road he lived on in Upper Norwood, London.[13] In contrast, the motor agent and aviator Toby Rawlinson (1867–1934) was born into higher circles. Eton and Sandhurst-educated, of an aristocratic family associated with the East India Company, Rawlinson resigned his commission as a polo-playing soldier in his twenties to immerse himself in the motor-traction business, notably with Darracq. Through his offices, the aviator and playboy Maurice Egerton (1874–1958) of Tatton Hall in Cheshire acquired a large Darracq motor car of his own.[14]

All were motor-traction entrepreneurs, for which there is no one single definition, other than an active engagement in the movement and promotion of its merits. As has been shown in the previous chapter, they operated in an environment where there was no market for motor cars and there was wide indifference amongst the public. 'Entrepreneur', then, could mean cold-calling salesmen such as Richardson, or gentlemen-salesmen such as Rawlinson. Richardson worked for a living while Rawlinson was a socialite, yet both were entrepreneurial in finding customers. In short, Rawlinson was the clubbable one, Richardson was not.

Edge's fit here is different again. He made a lot of money from motor traction and was a professional salesman by training, having moved from bicycle racing to sales. Yet he was able to move in elevated circles, mixing with royalty and aristocracy. He was the masculine hero of his age, competing in motor sport with notable success. He was a founder member of the Automobile Club, yet in the capacity of 'maker's amateur'. He became a gentleman-farmer but retained his connections and influence in the business of cycling and motoring.

The entrepreneur had the energy, contacts, initiative and access to funds (or probably, access to the people with the funds) to present automobility to the public and offer it as a consumer choice. It could include the cold-callers and gentlemen-salesmen described, just as it could 'fixers' like Claude Johnson, who facilitated the Rolls and Royce connection; or Charles McRobie Turrell, secretary to the British Motor Syndicate in the 1890s and right-hand man to Harry Lawson.[15] It could include journalists like Henry Sturmey, who edited *The Cyclist* (1877–1901) and the very first motor magazine, *The Autocar* (1895–1901); or Edmund Dangerfield (1864–1938), founder editor of cycling, motor-cycling and motoring magazines such as *Cycling* and *The Motor*. It could include the multilingual businessmen such as Frederick Simms, H. O. Duncan or Ernest

---

[12] Edge, 'Looking back: Interesting comparisons of yesterday and today by a pioneer', pp. 7–9.
[13] *Who's Who in the Motor Trade* (1934), p. 42; private correspondence with Penge Library.
[14] For Rawlinson see John Dyson, 'The adventurous life of Sir Alfred "Toby" Rawlinson', *Aspects of Motoring History*, 9 (2013), pp. 41–58. In 1904 Egerton acquired the 90hp Panhard used to break the flying kilometre record and converted it to a powerful tourer with a Rothschild body: *The Autocar*, 16 April 1904, p. 527. My thanks to Corey Estensen for this latter reference.
[15] Duncan, *The World On Wheels*, ii, p. 676.

Instone (1872–1932) who set up or staffed agencies abroad. It could also include those of working-class stock, notably Charles Jarrott (1877–1944), but who could use their skills, force of personality and achievements in the glamorous world of motor racing to operate easily in elevated social circles. They were all interdependent, and all 'necessary': the very wealthy might have found it more fitting to deal with someone from their club, for example, rather than visit agents' showrooms. These were the sorts who could see no contradiction between 'racing on Sunday, selling on Monday'.

This chapter, then offers a wide definition for the motor-traction entrepreneur but suggests some commonalities; it is especially concerned with the experience of S. F. Edge. Cycling was so often a common link, usually as an active club member, and often in a competitive capacity. In the pursuit of sporting or commercial goals they usually worked long and hard hours. They often needed tenacity, with a determination to work out technology, to tolerate unreliability and learn from it. They were image conscious, understanding that they needed to be seen as well dressed, to know when to take advantage of racing successes or new contacts and convert that into social capital to further develop their business interests. They knew when to parade wealth and success, yet appear as 'men of the people'. They all needed to be informed, to have business 'savvy'. Their experiences had often meant they had become accustomed to a good income in sales or competition, possibly as a 'maker's amateur', and this meant they would probably already have known minor fame. While so often they were men, women also made a clear impression as promoters of motor traction. Finally, Edge's political leaning was clearly Conservative, and this coloured his activities and pronouncements in the trade.

Cycling background

Edmund Dangerfield came to prominence in the 1890s when he launched the irreverent *Cycling* magazine. Brought up in a middle-class suburb of south London, he joined the Bath Road Cycling Club and Catford Cycling Club in the 1880s, and, on a 'safety', won several 100-mile races in 1890. At the time of the magazine's launch there were already twelve cycling magazines, indicative of the keen following that cycling as a sport and pastime brought, but also showing the gap in the market for a less stuffy read.[16] Dangerfield again identified a gap in 1902 when he believed the motor-cycle was the coming thing and launched *Motorcycling and Motoring*. That magazine published a feature, 'Who's who in the Light Car World', identifying seventeen men. Fifteen of those were based in the United Kingdom, and where they were associated with a brand, their names are recorded here, and while some of them recur in these pages, they were all entrepreneurs in the new motor-traction movement:

- S. F. Edge (of De Dion Bouton 'and other concerns connected with the motor trade')
- J. H. Adams (of Germain, manufacturer of motor cars)

---

[16] Armstrong, *Bouverie Street to Bowling Green Lane*, pp. 11–12.

- F. R. Goodwin (manager of the London depot, Star, cycles and cars)
- W. Williamson (managing director, Rex, motor-cycles and cars)
- S. D. Begbie (manager, Century, cars)
- H. T. Arnott (Princeps, motor-cycles and cars)
- G. W. Houk ('a typical Yankee, smart as they make 'em', Prescott, cars)
- H. Wait (manager, Clyde, cycles and cars)
- J. W. Stocks (manager, De Dion Bouton, motor-tricycles and cars)
- C. Friswell (Peugeot, cycles and cars)
- R. Burns (manager, Swift, cycles and cars)
- W. Munn (secretary, De Dion Bouton)
- H. H. Sturmey (Duryea, cars)
- W. M. Letts (Locomobile, cars)
- A. Burgess (secretary, MMC, cars)[17]

The report made the observation that they had made their names in the sport and trade of cycling. Some years later, in 1917, another magazine, *Motor Trader* reported on the annual dinner of the SMMT. Of the 177 diners it observed, 'one could not help being struck by the large number of old-time cyclists – particularly racing men – who have gravitated to "our" trade'.[18]

As the magazines pointed out, a remarkable number of entrepreneurs in motor-traction had arrived there through the sport and cultures of cycling. Many had had some experience of the cycling club, springing up all over the country from the 1870s onwards.[19] Club members tended to be self-selecting – middle-class, physically fit and male – although increasingly clubs became accessible to women and to couples on tricycles or tandems. Some of the clubs became prestigious or renowned, such as the Catford Cycling Club, the Anerley Bicycle Club (ABC) and the Bath Road Club. A number of motoring entrepreneurs were active in cycling clubs as young men in the 1880s and 1890s. Many cycling clubmen were members of multiple clubs, though where they were active, and prominent, as racers, tended to be associated with just one.

The cycle club often served as an 'apprenticeship'. Edge was associated most prominently with the ABC (he was active in others, for example, the Bath Road Club), and had joined it in about 1884.[20] The club was geographically convenient for where he lived in south London, and at one point he was joined on its weekly rides by three other members of his family, all male: his father Alexander (1843–1924), and brothers Kelburne and Seaton (1872–1910).[21] His first wife-to-be Eleanor (née Sharp [1872–1914]), living on the same street as the Edges, was also active in the ABC, and while

---

[17] *Motorcycling and Motoring*, 3 December 1902, p. 310.
[18] *Motor Trader*, 24 October 1917.
[19] See Ritchie, *Early Bicycles and the Quest for Speed*; Lawson, 'Wheels within wheels', pp. 123–45. See also, for example, Southcott, *The First Fifty Years of the Catford Cycling Club*; T. M. Barlow and J. Fletcher (comps.), *A History of Manchester Wheelers' Club, 1883–1983* (Manchester: 1983); [anon.], *The Black Anfielders Being the Story of the Anfield Bicycle Club, 1879–1955* (Anfield Bicycle Club, 1956).
[20] See Horner, 'S. F. Edge: The salad days', pp. 43–54.
[21] See the photograph of the ABC run taken at the Crown, Selsey 'in early nineties', which included all four Edge men: Smith, *Some Notes about the Anerley B.C.*, p. 4.

there were one or two other women mentioned in club accounts, she was unusual in actually venturing out on a bicycle on club rides.[22]

Edge was also active socially and on the ABC's committee, serving as its secretary, a job that required some initiative in organization and delegation as well as in keeping the cycling press informed. As was typical, the ABC had a schedule of races and time-trials on its calendar, which involved challenging other clubs. These events sometimes lasted days and required lengthy travelling, and it would have been necessary to organize pace-riders; accommodation; starting points and itineraries; timing officials; alerting journalists from the cycling press; routes; food; and support.[23] Depending on the strengths of its members, clubs would often be associated with particular types of competition, such as the 100-miler, or the 24-hour. It was while representing the club that Edge set several cycle records in the 1880s and 1890s; his first newsworthy triumph had been in 1888 when he broke the London-Brighton-London record. Here, the interplay of sports (and the likely cross-pollination of the same people active in different sports) meant that horse-driving feats became challenges for cyclists, and this had been the impetus for Edge. Earlier that year coachman James Selby had set the record for the same route driving 'Old Times', a coach-and-four, requiring fifteen changes of horses in an operation so slick that one change took just 47 seconds.[24] For Edge's challenge, on largely unsealed roads, on a single-speed bicycle with 'cushion' tyres, he beat Selby, setting the record at 7 hours and 2 minutes. He went on to set a London-York cycling record in 1892 (12 hours, 49 minutes) with the same methods.[25] Achieving these records was also a team effort, and each contender benefited from an interdependent spirit of discipline and self-reliance.

The club provided the impetus for competition and the variety of challenges meant a broad range of cyclists could participate – that is, the sprints, the endurance, the long distance, and the recent availability of the 'safety' bicycle. A fanatical following and a very active cycling press meant that record breakers became household names. They were known about nationally and welcomed whenever they competed around the country. They were used to draw in larger, paying crowds of spectators. Here, cycle track races were often staged by entrepreneurial publicans on land adjoining their premises, and interest in the sport is clear in how it continued to attract large, paying crowds, well into the 1890s.[26] For the 24-hour Cuca Cocoa Challenge Cup race at Crystal Palace in 1892, when cycling 'crack' (famous racer) Frank Shorland beat Arthur V. Linton (c. 1872–96),[27] 15,000 spectators were there for the finish but, even more impressively, over one thousand had stayed to watch overnight when the bicycles used 'head-lights'

---

[22] Unpublished diary (1891–6) of G. H. Smith.
[23] Described in Smith, *Selwyn Francis Edge*.
[24] See Charles G. Harper, *The Brighton Road: Speed, Sport and History on the Classic Highway* (London: Chapman and Hall Ltd, 1906), pp. 157–63; Viscount Bury and G. Lacy Hillier, *Cycling* (London: Longmans, Green, & Co, 3rd ed., 1891), pp. 212–5, 456–9.
[25] Smith, *Some Notes about the Anerley B.C.*, p. 30; Smith, *Selwyn Francis Edge*, pp. 17–19.
[26] Samantha-Jayne Oldfield has described the role of the publican in facilitating sporting events in the nineteenth-century: see her 'Running pedestrianism in Victorian Manchester', *Sport in History*, 34:2 (2014), pp. 223–48.
[27] https://gracesguide.co.uk/Francis_William_Shorland; https://gracesguide.co.uk/Arthur_Linton all accessed 20 June 2019.

(although it is unclear what exactly the crowd would actually have seen).[28] Shorland, in fact, had been the first to beat Selby's London-Brighton-London record in 1888, before Edge eclipsed that a few weeks later. Subsequently Shorland moved through a series of sales positions for cycle companies, ultimately becoming general manager for the motor-car manufacturer Clement-Talbot.

Cycling clubs also provided dinners and socials, particularly during the winter months when roads were especially difficult to navigate. Social functions provided an opportunity for networking, but also served as a primer for the clubs' young members in terms of etiquette, particularly useful when they sought membership in later years in more expensive and mannered gentlemen's clubs such as the Automobile Club. Descriptions of the socials of the cycling clubs, meanwhile, offer a window into the changing patterns of behaviour for the middle-class sportsman at ease: dances, for example, went out of fashion 'with the rather unsociable character modern dances and dancers have assumed'. In 1893, eighty-seven members and 'their lady friends' 'tripped the night hours away'. Regular events included billiards, card competitions, concerts, lantern shows, hot-pot suppers, whist drives and 'home trainer competitions' which were 'extremely exhausting', 'none the less so in a hot and smoke-filled room'.[29] As the summer runs closed, 'after tea the ring was cleared and all sorts of sports and fun were indulged in, boxing, cock-fighting, wrestling', as well as singing.[30]

G. H. Smith was active in the ABC in the 1890s. He became an assistant editor for *Cycling*; a sales representative for Dunlop tyres; in 1900 general manager for United Motor Industries (based in Paris, with Duncan as managing director); and honorary secretary of the Motor Trades Association (with Harry Lawson as its president).[31] Smith kept a diary of ABC club activities which described the dreadful roads, the abuse from bystanders and the calamities, but also the camaraderie and exclusive society that amateur club cycling brought. He also described a strict cycling-club hierarchy which privileged the position of president, who was key in the success of the club's events. Not normally expected to take any active part in the club, the president served more as a 'patron'.[32] ABC member and bicycle racer Harry Swindley (c. 1861–1918) believed the president constituted the 'esprit de corps' in a cycling club, requiring a 'generous and sporting' personality; Swindley in turn became a cycling journalist and a member of the council of the SMMT. Smith described the club dinners of the early days, initially at the Bridge House Hotel, Southwark, moving in 1893 to the Holborn Restaurant, and the part the president played in these. At that time the president of the ABC was the

---

[28] Smith, *Some Notes about the Anerley B.C.*; E. K. S. Rae, 'Unconventional portrait of the week: Mr Percy Richardson', *Automotor Journal*, 2 April 1910, pp. 355–7.
[29] Smith, *Some Notes about the Anerley B.C.*, pp. 47–8.
[30] Smith, *Some Notes about the Anerley B.C.*, p. 14 and esp. 'Festivities and Customs'. See also diary of G. H. Smith. See also University of Warwick, Modern Records Centre, for reports of the dinners of the North Road and Stanley Cycling Clubs: MSS.328/N7/1/94. For the ritualistic carrying off of the cup-winner at the 1896 Stanley Club dinner, indicating the shift from formal to informal: '*Sketch* Cycling Supplement', *The Sketch*, 29 November 1896, cover.
[31] https://gracesguide.co.uk/G._H._Smith, accessed 21 June 2019. See also Armstrong, *Bouverie Street to Bowling Green Lane*, p. 32.
[32] *The Black Anfielders*, p. 1.

businessman and bookmaker R. H. Fry (1836–1902), the 'perfect chairman and ruler of the feast', and who paid for wines, cigars, waiters' tips and most of the prizes.[33]

The position of president was intended for men of prominence in business or the community, usually older men who had made their mark. It served as a marker of status to be offered the position. Harry Lawson had been president of Watford Cycle Club in 1893.[34] Edge had stayed in touch with the ABC once he moved into the motor-traction world and was in turn elected president for 1929–30; he had also been president of the Catford Cycling Club for 1926–7.[35] He would have been an excellent catch in this role, by then in his dotage and a household name as a successful, if long retired, racing driver. He in turn had been president of the SMMT (1913–14) and prior to this, its vice-president four times and treasurer once.[36] The endowing of trophies and awards served to honour and perpetuate the memory of the endower, and was also indicative of the sense of kinship that club membership brought. Edge had already sponsored the S. F. Edge Trophy in 1903, a 200-mile non-stop motor-cycle trial for the Motor Cycle Club.[37] He then sponsored ABC events, adding the Old Boys' Section from 1905 and the S. F. Edge Novice's Cup. His Old Boys' Cup had become, by 1928, a day out on a steamer, known as 'Edge's Run', with his old club friends in the ABC. From the late 1920s he entertained the ABC on their runs at his estate in Sussex.[38]

Graft and learning

Cycling as a sport was physically and mentally demanding. Track and road events required stamina, technique and fitness, while even club rides demanded concentration and tough character. Cycling was generally ill-suited for the roads, many out of towns having loose surfaces, mud or awkward camber. The 'ordinary' rider was liable to a 'header' (going over the handlebars) while all cyclists could have a 'cropper' (falling off). Any breakages or punctures on the road required initiative and often, strength, to repair, to avoid wheeling the bicycle to the nearest train station to return home. The cycling fraternity, and a national cycling press, encouraged a knowledge base to be built up. Tricks and work-arounds at the roadside were reported in the magazines and such advice assisted newcomers to the sport. The cycling club provided a forum for local knowledge. The same cultures informed the first motorcyclists and motor-car drivers, assisted by the rapid appearance of motoring clubs and a specialist press.

---

[33] Smith, *Some Notes about the Anerley B.C.*, pp. 43–4, 50. For Richard Henry Fry, see Alun Thomas, 'The leviathan of the turf: R.H. Fry and the Grecian Villa', *Norwood Review*, Autumn 2017, pp. 11–18. The second president was Sir John Blundell Maple, MP (Smith, *Some Notes about the Anerley B.C.*, p. 50), underlining how the president needed to be highly connected and generous with his money.
[34] Duncan, *The World On Wheels*, ii, pp. 671.
[35] Smith, *Some Notes About the Anerley B.C.*, p. 51; Southcott, *The First Fifty Years of the Catford Cycling Club*, p. 476.
[36] Nixon, *The Story of the S.M.M.T., 1902–1952*.
[37] *The Motor Cycle*, 27 May 1903, p. 194.
[38] Smith, *Some Notes about the Anerley B.C.*, pp. 12, 17, 58.

It was necessary, then, for cyclists and then motorists to be willing to 'muck in' to build up personal experience and knowledge. There were no user guides or manuals to consult, and even maps were often out of date. For example, pioneer motorist T. R. B. Elliot (c. 1871–1949), who claimed to be the first motorist in Scotland,[39] described the arrival in December 1895 of his first car, a Panhard and Levassor 3½hp phaeton, in a large wooden case, from Paris. 'A German, who knew very little about the car, arrived the same day, in order to give me the "points" of the new vehicle', Elliot recalled. Once unpacked, the visitor was unable to start the car, explaining, '"the sea air must have got into the motor"'. Elliot in the end gathered ten men who after four hours managed to start the car.[40]

Edge has described in his autobiography his formative experiences with the earliest motor-cycles and motor cars. He had met Lawson in 1893 when he, Edge, was general manager for the Rudge cycle company. By 1896 he approached Lawson again, and this approach coincided with one from Charles Jarrott, a young sportsman then engaged with a firm of solicitors.[41] Both young men had recognized the need to learn the workings of motor vehicles from first principles, and Lawson was happy for them to take apart, reassemble and test-drive the many motor cars and motor-cycles at his disposal. For Edge, this experience was conducted in his spare time (8pm to 8am) and during weekends. He related how he and Jarrott emerged from Lawson's grand premises on Holborn Viaduct in the small hours one day, 'covered in grime and looking like a couple of dishevelled deep-sea fishermen' and were accordingly refused admission to a restaurant.[42] Similar stories of potential social embarrassment were related by others, for example T. R. B. Elliot, who having taken nine hours to cover the fifty miles to Edinburgh in 1896 was 'too dirty to go to my club', so stopped at a hotel.[43]

The necessity of practical knowledge at a time when there were so few motor vehicles in the country, was not lost on others, and social class need not have been a barrier to gaining knowledge. At one extreme, Major George Cornwallis-West (1871–1951), for example, wanting to know more about the engine in his De Dion, took a job in a Paris garage under the name of Smith.[44] At the other, Jarrott, like Edge, secured his status as racing driver and motor-traction entrepreneur through skill but also force of personality. This latter point was important; as the son of a blacksmith's labourer, he would ordinarily have had no place in the Automobile Club. His aping the manners 'of more fortunate people', and assuming a 'gentlemanlike accent' apparently only held 'while he remained moderately sober', something which Jarrott in later years found increasingly difficult to do.[45]

---

[39] See https://gracesguide.co.uk/Thomas_Robert_Barnewall_Elliot, accessed 20 June 2019.
[40] T. R. B. Elliot in Montagu, *The First Ten Years of Automobilism, 1896–1906*, p. 18. For an idea of the choreography of starting a veteran motor car, see https://www.prewarcar.com/pre-war-of-the-week-an-1894-peugeot-on-hottube-ignition-how-to-start-and-run-this-piece-of-history, accessed 18 July 2019.
[41] https://doi-org.mmu.idm.oclc.org/10.1093/ref:odnb/64993, accessed 17 April 2020.
[42] Edge, *My Motoring Reminiscences*, esp. pp. 21–3.
[43] T. R. B. Elliot in Montagu, *The First Ten Years of Automobilism*, p. 20.
[44] Brendon, *The Motoring Century*, p. 40.
[45] The son of a blacksmith's labourer. See Elizabeth Ellen Bennett, 'Jarrott, Charles (1877–1944)', *Oxford Dictionary of National Biography*, Oxford University Press, 2004; online edn, Sept 2014

Figure 3.2 S. F. Edge, assisted by (probably) Brooks, repairs a puncture at the roadside on the Napier during the Thousand Mile Trial. They are on the road between Darlington and Northallerton, it is 7 May 1900, and St John Nixon and (probably) Mary Kennard look on. Image: National Motor Museum, Beaulieu.

Edge took care to be informed of all matters relating to the bicycle and motor trades. According to Duncan, he took hours out to read all the 'motor papers carefully', and in conversation he would 'know that' when friends referred to a wide variety of topics.[46] He made notes about the business people he met, and rated them as to whether he could do business with them.[47] He described deriving satisfaction in studying the designs of the motor vehicles and rectifying their faults; some twenty-seven patents can be traced to his name relating to motor-vehicle improvements.[48] He made his first unaccompanied motor trip, to Canterbury, in 1897 on Lawson's De Dion motor tricycle, and worked out improvements that could be made. Edge showed particular initiative in purchasing the 6hp Panhard-Levassor motor car (known as Number 8) which had won the 1896 Paris-Marseilles race. Lawson had already arranged to purchase it to ensure it was paraded in his 1896 'Emancipation Run', and Edge subsequently used it as a development mule, teaming up with former cycling companion Montague Napier, who had inherited the family precision-engineering company, to effect the improvements.[49]

[http://www.oxforddnb.com/view/article/64993, accessed 8 July 2016]. 'Aping': Brendon, *The Motoring Century*, p. 39. Letter, S. C. H. 'Sammy' Davis to Ronald 'Steady' Barker, 3 December [1969?], provided courtesy of Jonathan Rishton.

[46] Duncan, *The World On Wheels*, ii, p. 847.
[47] Edge scrapbooks, vol. 141, Veteran Car Club archive.
[48] Conducting a search of espacenet.com for 'Selwyn Francis Edge', 1895–1940, accessed 20 October 2019.
[49] Edge, *My Motoring Reminiscences*, pp. 40–4; Vessey, *By Precision into Power*, pp. 47–9; Duncan, *The World On Wheels*, ii, p. 808–11, 828. Edge is photographed, wearing a bowler hat, aboard the unmodified Panhard: *Veteran and Vintage*, 4:3 (1959), p. 452.

Figure 3.3 S. F. Edge imported De Dion motor-tricyles and cars. Here he is on a 1901 tricycle with an air-cooled 2¾hp engine. This model is unusual in that it is not water-cooled and it has no gearbox. Edge has evidently specified equipment here for long-distance travel, such as a good-quality front lamp, horn, large-capacity fuel tank, and tool roll. The same photograph appears in Selwyn Francis Edge, *My motoring reminiscences* (London: G.T. Foulis, 1934), opp. p. 48, in which Edge incorrectly captions it as a 6hp racing machine. Image: Malcolm Jeal, with identification and correction thanks to Michael Edwards.

The celebrity of the cycling 'crack'

The image of the cycling 'cracks' in the 1880s and 1890s were in part created and then built up by the cycling magazines. Interviews and features were an important component of each magazine's regular copy. This helped build up a celebrity status, used by promoters to draw in larger and paying crowds for track events throughout the country. The 'crack' in turn benefited through higher remuneration: *Punch* in 1896 mentioned that 'one well-known wheeler [cyclist] has earned £2,000 in prizes alone' in seven months, in addition to the 'heavy retainer' from the tyre and cycle manufacturers whom he represented. 'A leading professional cyclist', *Punch* continued, had an income from £1,500 to £2,000 a year.[50] These cyclists were not identified, but the report gives a measure of their potential income.

A number of these cycling heroes went on to become agents for cycling brands, and then motoring – they became the motor-traction entrepreneurs. Edge was one of

---

[50] 'Calves and cash', *Punch*, 5 September 1896, p. 117.

these, and he took care to cultivate the new image of the strong, masculine Edwardian hero. As described by Tosh and others, Edge fitted the late-nineteenth-century image of manliness and physical endeavour, standing out as an exemplar of the handsome, manly and healthy amateur sportsman of the age.[51] Boutle has shown how setting speed records in the period to the First World War helped project an image of a 'modern, dynamic nation';[52] Edge went on to set a non-stop 24-hour driving record on the newly-opened Brooklands circuit in 1907. The strident opinions of 'Owen John' in *The Autocar* in 1914 echoed a world with which Edge was associated:

> Motoring has effected many changes, but slovenliness is not one of them, rather it should be thanked for getting rid of the 'horsey' type that was once so painful to gaze on and had such a host of imitators. As a matter of fact, motoring – in spite of some drawbacks that have come with it – has done a great service to society by eliminating the long-haired type of effeminate non-descript that once upon a time endeavoured to set the fashion. Nowadays long hair and manners en suite are not suitable for motoring, and just as the South African War [1899–1902] was responsible for the fashion of a military appearance, so a knowledge of mechanism has effected a more manly type of youth.[53]

Edge and his peers needed to be fit to meet the physical demands of, first, cycle and then motor-cycle time trials, and to handle the brutish racing cars of the early 1900s. Clean-living was also part of the image. In an interview in 1895, Edge revealed that his methods included training with dumbbells and eating three good meals every day.[54] He never smoked and he enjoyed a moderate diet with water, light cider or claret.[55] He embraced and endorsed the physical culture and regime of the showman Eugen Sandow (1867–1925), a household name and idealized as an icon of manly perfection and epitome of health. That Sandow in turn endorsed Edge, calling him 'the successful athlete, motorist and business man', is a useful marker of the esteem in which Edge was held.[56] During the preparation for the 1903 Gordon Bennett race in Ireland, Edge, in

---

[51] Philip Gordon Mackintosh and Glen Norcliffe, 'Men, women and the bicycle: Gender and the social geography of cycling in the late nineteenth century', in Dave Horton, Paul Rosen and Peter Cox (eds), *Cycling and Society* (Aldershot: Ashgate, 2011), here pp. 155–6. See, for example, John Tosh, 'Masculinities in an industrialising society: Britain, 1800–1914', *Journal of British Studies*, 44 (2005), pp. 330–42; John Tosh, *Manliness and Masculinities in Nineteenth-Century Britain: Essays on Gender, Family and Empire* (Harlow: Pearson, 2005). Also useful for contextualization is Graham Dawson, *Soldier Heroes: British Adventure, Empire and the Imagining of Masculinities* (London: Routledge, 1994), part II. For entrepreneurialism in amateur sport, see Wray Vamplew, 'Sport, industry and industrial sport in Britain before 1914: Review and revision', *Sport in Society*, 19:3 (2016), pp. 340–55; and Porter, 'Entrepreneurship', pp. 197–215.
[52] Ian Boutle, '"Speed lies in the lap of the English": Motor records, masculinity and the nation 1907–14', *Twentieth-Century British History*, 23:4 (2012), pp. 449–72 at pp. 452–4, 459.
[53] 'Owen John', 'On the Road', *The Autocar*, 25 July 1914, pp. 170–2.
[54] 'How I got strong: revelations of a strong man', *Chips Illustrated*, 16 February 1895.
[55] Duncan, *The World On Wheels*, ii, p. 850.
[56] Christopher Breward, *The Hidden Consumer: Masculinities, Fashion and City Life, 1860–1914* (Manchester: Manchester University Press, 1999), pp. 242–6. See also Mark Pottle, 'Sandow, Eugen (1867–1925)', ODNB, https://doi-org.ezproxy.mmu.ac.uk/10.1093/ref:odnb/76284, accessed 20 August 2019. For endorsements: *Daily Mail*, 26 April 1909, Sandow ad.

contrast with his more relaxed co-drivers, 'indulged in Sandow's exercises' to maintain his fitness while waiting for the race to start.[57] Edge was the all-round sportsman, a keen swimmer, excellent shot, a yachtsman, as well as a boxer, with contacts to arrange boxing tournaments for the ABC.[58] When Edge had completed his Brooklands 24-hour non-stop drive in 1907, he sat for interviews with the motoring press[59] to play up the image of health and manliness. This particular event, while a clear triumph for Edge and the Napier brand, was stage-managed: Edge paid one hundred guineas to be the first person to use the brand-new circuit,[60] and it was then a model of teamwork to facilitate his drive. While Edge went to some trouble to be seen as clean-living, others in a similar position were more relaxed. Jarrott, for instance, was often photographed with a cigarette in his mouth, and played cricket as Edge worked out. Whatever image these two men worked on, they both had to be fit to withstand the rigorous demands of motor racing, for which they both became famed.

Here, Edge was tipped into the realm of a 'household name' when he won the 1902 Gordon Bennett Cup, an international motor race held annually between 1900 and 1905; that year it went from Paris to Innsbruck. Edge was one of the few entrepreneurs representing Britain to commit to international competition, in his case sponsoring the Napier brand (others included Herbert Austin [1866–1941], then working for Wolseley).[61] Stories circulating at the time, and repeated in Edge's autobiography, describe one particular feat during the race – the changing of covers (tyres) by hand, even though he had lost his tools and jack.[62] When St John Nixon was ghost-writing Edge's autobiography in 1934, he asked Edge for clarification, puzzled why Edge found it necessary to change the covers when he had already been informed he had won the race, all other contenders having dropped out. Nixon asked how it was possible to change the tyres without tools: 'I should have thought this would have been practically impossible'. Edge scribbled in reply, 'So would I if we had not done it'.[63] These were the stories which helped build up the image of the hero. Indeed, Jarrott told a story of his own in his 1906 autobiography. He had been participating in the concurrent Paris-Vienna race in 1902, covering the same ground as the Gordon Bennett at the same time, and he was in Bregenz on the same night as Edge. Jarrott described cutting up the wooden bed in his hotel room to use to shore up the stricken chassis on his Panhard.[64] This particular race, the 1902 Gordon Bennett, was covered in the national press and provided a fillip for the domestic motor industry – Duncan wrote, 'no event [winning

---

[57] Duncan, *The World On Wheels*, ii, p. 850; Edge, *My Motoring Reminiscences*, pp. 144–5.
[58] Smith, *Selwyn Francis Edge*, p. 6; Duncan, *The World On Wheels*, ii, pp. 842, 850; Smith, *Some Notes about the Anerley B.C.*, p. 47.
[59] For an account of the feat see Venables, *Napier: The First To Wear the Green*, pp. 92–4. See also Boutle, '"Speed lies in the lap of the English"', p. 457.
[60] *Daily Mail*, 1 January 1907.
[61] W. F. Bradley, *Motor Racing Memories* (London: Motor Racing Publications, 1960), pp. 28–34. See also [Lord] Montagu of Beaulieu and Michael Sedgwick, *The Gordon Bennett Races* (London: Cassell, 1965); Gerald Rose, *A Record of Motor Racing 1894–1908* (1909; London: Motor Racing Publications, 1949), chaps. 7–12.
[62] Edge, *My Motoring Reminiscences*, pp. 108–24.
[63] Nixon/1, loose papers, National Motor Museum library, Beaulieu.
[64] Jarrott, *Ten Years of Motors and Motor Racing*, pp. 137–41.

the Gordon Bennett] in the history of automobile sport or trade had such a wonderful effect upon the motor industry in England'.[65]

The entrepreneur also needed to understand, and manipulate, an image of glamour. Edge was conscious of how he was presented in the press, as his collection of newspaper clippings, assiduously gathered by his second wife Myra Caroline (née Martin) (1887–1969), testifies.[66] Hahn has pointed out how the Victorians were 'fascinated with celebrities like nowhere else, whether writers, military heroes, explorers, scientists, artists, members of the high society, actors or spiritual mediums'.[67] The motoring promoter Edward J. Pennington had a diamond-studded gold watch which, when in the company of prospects, he affected to casually extract from his waistcoat as though by accident. His wife was used to draw in the wealthy set: for example, at the 1896 Motor Car Club show at the Imperial Institute, she positioned herself prominently, fashionably dressed and draped in diamonds.[68] Clothing and fashion mattered to give a good impression in business and social circles. According to Breward, by the 1890s male models in the sartorial press were younger, more athletic than hitherto, and were wearing lounge suits, all to the benefit of younger entrepreneurs like Edge who could take advantage of the more casualized code of dress that cycling first brought.[69] Edge understood, like Lawson, Hooley and businessman Arthur Du Cros (1871–1955), the necessity of 'looking' the part to instil confidence and credibility in his investors and customers. His many sessions posing for photographers show a well-dressed, fashion-conscious, confident man.[70] Like Charles Friswell, who promoted his 'Automobile Palace' at 48 Holborn Viaduct, a five-storey building with every need of his customers catered for, Edge had his New Burlington Street premises in London, opened in 1902, with plate-glass windows. It had room for 150 cars, and Napier and Gladiator motor cars were sold there. The premises were considered, in 1907, 'the most elegant in Europe'.[71] At that time Edge had a 'splendid suite of apartments' at Whitehall Court. With a view onto the Thames, it had a full-sized billiard table on the upper floor, with each lower room finely furnished and 'gaily decorated with plants and flowers'. Edge became very wealthy through his business ventures, particularly so when he severed his business connection with Napier in 1911.[72]

---

[65] Duncan, *The World On Wheels*, ii, p. 838.
[66] The collection is now held at the Veteran Car Club archive.
[67] H. Hazel Hahn, 'Consumer culture and advertising', in Michael Saler (ed.), *The Fin-de-siècle World* (Abingdon, New York: Routledge, 2015), p. 396. See also Simon Morgan, 'Celebrity: academic "pseudo-event" or a useful concept for historians?', *Cultural and Social History*, 8 (2011), pp. 95–114.
[68] Duncan, *The World On Wheels*, ii, pp. 681–8, 775.
[69] Breward, *The Hidden Consumer*, pp. 39–41.
[70] Arthur Du Cros, like his father, was known as the 'best-dressed man' in the House of Commons, and was always immaculately presented: G. K. S. Hamilton-Edwards, 'Du Cros, Sir Arthur Philip, first baronet (1871–1955)', rev. Geoffrey Jones, *Oxford Dictionary of National Biography*, Oxford University Press, 2004; online edn, Oct 2009 [http://www.oxforddnb.com/view/article/32914, accessed 19 July 2016].
[71] *Automotor and Horseless Vehicle Journal*, 2 August 1902; Morrison and Minnis, *Carscapes*, esp. chap 2: 'Selling cars', and pp. 47–57 for a discussion of premises, architecture etc (p. 52 for discussion of Edge's premises).
[72] Edge was to receive £25,000 in cash and £125,000 in shares and securities: *Daily Mail*, 10 December 1913.

Travel abroad fitted this image of glamour and success. Edge and Jarrott described trips to the United States, South Africa[73] and Australia, and both had visited and raced in continental Europe since the 1880s. Edge in later life claimed to have completed a lengthy continental trip every year since 1900, always in a British car.[74] Edge's family was Australian – he had been brought over to the United Kingdom aged three – and his father and siblings frequently made journeys between the two countries. The image of glamour and success was reinforced by the cycling and motoring press which reported on all grand dinners and social events. Lawson had a knack of ensuring the press were on hand whenever he met royals, aristocrats or parliamentarians. For his Motor Car Club show at the Imperial Institute, from May to August 1896, Lawson arranged for posters to be affixed all over the capital, attracting the Prince of Wales, MPs and minor royalty.[75] In his turn, Edge was able to secure an exclusive personal tour of motor shows with Prime Minister Balfour (1848–1930), a keen cyclist and motorist,[76] who had bought numerous Napiers in 1902 and 1903.[77] Edge then provided Balfour with transport for his visit to the Olympia Motor Show in 1905,[78] when Edge showed him around the exhibition, accompanied by Lords Salisbury and Lansdowne.

Putting these elements together, the entrepreneur was able to work on creating a brand. For Edge this was Napier – he raced them and he had sole rights to selling them. Edge placed adverts for Napier cars in the national, sporting and country-set press, as well as the motoring magazines. By 1903 his Napier catalogue was offering a three-month guarantee on material and workmanship, and the claim: 'If on delivery, it does not fulfil the claims made for it, the purchase money will be refunded in full, with interest'.[79] Journalists could expect dinners and entertainments whenever a grand announcement was made, for example, in 1906 when his company S. F. Edge Ltd, the agency for Napier, began to offer a three-year guarantee on every Napier. Furthermore, the company was now guaranteeing to pay twenty pounds per week for failure to deliver on time.[80] (This guarantee was subsequently challenged by a disgruntled customer Asher Wertheimer (1843–1918), a celebrated art dealer of Bond Street, who disputed the worth of the three-year guarantee on two Napier cars which he had found unsatisfactory; the case for £120 was won by Edge when he was able to show that work had been undertaken by non-approved agencies.)[81] In addition, as the sole agent for Napier cars, Edge was able to demand of Napier high standards of quality control, but

---

[73] See also Duncan, *The World On Wheels*, ii, p. 847; and Bob Johnston and Derek Stuart-Findlay, *The Motorist's Paradise: An illustrated History of Early Motoring In and Around Cape Town* (Cape Town: Bob Johnston and Derek Stuart-Findlay, 2007).
[74] *The Times*, 29 September 1928.
[75] Duncan, *The World On Wheels*, ii, pp. 681–4.
[76] Balfour first bought a De Dion tricycle in 1899: see Paul Harris, *Life in a Scottish Country House: The Story of A.J. Balfour and Whittingehame House* (Whittingehame: Whittingehame House Publishers, 1989), pp. 89–91. In a letter to Lady Elcho, Aug. 1901, he reveals his motoring 'advisers' were at 14 Regent Street 'in the person of W. [sic] Jarrott': see Jane Ridley and Clayre Percy, *Letters of Arthur Balfour and Mary Elcho, 1885–1917* (London: Hamish Hamilton, 1992), pp. 180–1.
[77] *Daily Mail*, 31 January 1903.
[78] *The Times*, 13 February 1905.
[79] Napier brochure (1903), p. 12, with thanks to the late Malcolm Jeal.
[80] Duncan, *The World On Wheels*, ii, p. 846; *The Times*, 3 March 1906, ad.
[81] *The Times*, 18 December 1909; *Daily Mail*, 17 December 1909, 18 December 1909.

then also made sure the press was 'aware' of this. In return, Edge expected compliant copy. In 'Motor Matters', a regular column in *The Times*, for example, the journalist was impressed with the build process of every Napier. He noted that the rolling chassis were sent to S. F. Edge Ltd, tested, sent to the body builders, then tested for two hundred miles.[82] Edge clearly understood that to get the sale, he needed to keep – and be seen to keep – quality up.

Edge was not a particularly gripping writer – St John Nixon wrote his autobiography for him – but his name had become a marker of common sense and know-how, with contributions to publications as diverse as the Badminton Library (1902), and a book on the frontiersman (1909).[83] He guest-edited the *Penny Illustrated Paper*[84] and wrote forewords for books on motoring and travel.[85] His forte was the letter, and he wrote prolifically to the editors of the cycling and motoring magazines – forty-two in *The Autocar* in 1904 alone, for example,[86] and to the farming press after 1912 when he moved into pig-farming. Even once he no longer needed to, Edge continued to 'muck in' at events such as trials; he served as 'honorary marshall' for the 650-mile trial of the Automobile Club in 1902 (even though by this time he was the Gordon Bennett hero), which guaranteed his name and photograph were in the motoring magazines.[87] He also attached himself to specific charities, notably, in 1913, the Pure Food & Health Society, a pressure group for improving food standards, and through which milk was provided at cost for the poor in London.[88]

Background in business and sales

Family background was a factor for many who moved into the fledgling motor-traction business. For Edge, business began at home, a large detached property in the new south London suburb of West Norwood, convenient for the Crystal Palace. It was comfortably middle class – the family had four servants in the 1880s. His father Alexander, and Alexander's brother Henry (1845–1925) floated and managed a business importing meat from South America in the 1880s.[89] As the eldest son, Edge was left to take charge of the family interests when his now-widowed father emigrated to Australia in 1892 with Edge's brothers and sisters. Like many of his peers, Edge was

---

[82] 'Motor matters', *The Times*, 11 July 1907, in praise of Napier.
[83] S. F. Edge and Charles Jarrott, 'Motor driving', chap. in Alfred Harmsworth (ed.), *Motors and Motor Driving* (London: Longmans, Green & Co, 1902); S. F. Edge, 'Motor cars and modern warfare', in Roger Pocock, *The Frontiersman's Pocket-book* (1909; Edmonton, AB: University of Alberta Press, 2012), pp. 223–6.
[84] *Penny Illustrated Paper*, 8 February 1913, from Boutle, '"Speed lies in the lap of the English"', p. 463.
[85] The book, [Anon], *The British Motor Tourist A.B.C.* (London: The British motor tourist A.B.C. Co., 1905) is a case in point. Edge's foreword, 'Hints to motor tourists' (p. vii) is rather dull and, despite the anonymous editor's announcement, was clearly not 'specially written' for the book.
[86] J. M. Fenster, 'Mr Jarrott and Mr Edge: An Edwardian rivalry', *Automobile Quarterly*, 29:1 (1991), pp. 96–105.
[87] *Automotor and Horseless Vehicle Journal*, 6 September 1902, pp. 501, 506.
[88] *The Times*, 5 November 1913. See Michael French and Jim Phillips, *Cheated not Poisoned? Food Regulation in the United Kingdom, 1875–1938* (Manchester: Manchester University Press, 2009), p. 137.
[89] Horner, 'S. F. Edge: The salad days', p. 46.

willing to relocate for work – as a cycling salesman in the early 1890s he was posted to Coventry, Glasgow, and then back to London, taking his first wife Eleanor with him each time.[90] Edge was also in a position to observe and learn from those already active in business; he learnt, for example, when identifying and attracting his company directors, that they should be active and effective. Here, Harry Lawson's inclination to attract peers to his board, usually by paying them, thus putting glamour before ability, was one of his weaknesses.[91]

Edge's brother Kelburne had been active in the ABC up to the time when he emigrated to Australia in 1891 'to make use of those business capacities he ha[d] so often displayed in the club's [ABC's] service, and, we hope, to make his fortune'. There, Kelburne turned to whatever business he could –variously, stationer, electrician, importer of Yost typewriters and bicycle importer,[92] but also as an agent for the Austral Cycle Company, set up in 1896 with Edge in the United Kingdom and Kelburne in Australia, and the Du Cros brothers as directors. This turned out to be an inopportune moment to set up a bicycle business, and it failed in 1899, following overexpansion in the bicycle trade just as the market was bottoming out.[93] Edge used his name to endorse motoring and cycling accessories[94] but was also willing to invest beyond his cycling and motoring interests, including in the film industry,[95] dirigible airships[96] and bodybuilding regimes. He was also recognized as an expert in his field, and gave evidence to Royal Commissions, such as the one on Motor Cars in 1905 when he represented the SMMT. He was appointed to a Royal Commission in 1909 for an Exhibition of Arts.[97] He subsequently invested in pig-farming and was soon pontificating in the agricultural press with his customary energy and conviction.[98]

Entrepreneurial women

Few women were visible in cycling or motor-car sales and publishing ventures. Women had played no active part in horse-driving in the latter part of the nineteenth

---

[90] See Horner, 'S. F. Edge: The salad days', for the background of family members.
[91] For example, Duncan believed a reason why Lawson's British Motor Syndicate failed was that, 'Unfortunately he had a weakness for "dummy directors" on his boards': Duncan, *The World On Wheels*, ii, p. 733.
[92] *Cycling*, 24 January 1891. I am grateful to Rosemary Sharples for drawing my attention to Kelburne's business interests in Australia: see http://www.launcestonfamilyalbum.org.au/detail/1030873/kelburne-ernest-edge, accessed 31 May 2019.
[93] *Financial Times*, 8 May 1899, 20 May 1899.
[94] Many references can be found amongst the adverts, especially in the cycling and motoring press. The first found to which he put his name was 'Anti-Stiff', which he confirmed removed aches and pains: *Cycling*, 31 January 1891, ads. *The Book of Edge Accessories for Motor Cars* (1912) reveals the full extent of the power of his ability to endorse, here a catalogue full of Edge-branded lighting, speedometers, tool rolls, oils and greases (my thanks to the late Malcolm Jeal for this reference).
[95] David Fisher, *Cinema-by-Sea: Film and Cinema in Brighton and Hove since 1896* (Hove: Terra Media, 2012), pp. 159, 163, 187.
[96] *Daily Mail*, 4 October 1909.
[97] *The Times*, 21 October 1905, 24 March 1909.
[98] Including *Agricultural Gazette* and *Country Life*. He also appeared in editorials in the *Implement and Machinery Review*.

century, although there were female proprietors of staging inns who facilitated driving activities, and a few women drove horse-omnibuses. Articles in society magazines and, the heroines in novels, tended to place the female cyclist or motorist at the mercy of the male entrepreneur. One example is 'Janet Jenks', Mary Kennard's fictional character in her *The Motor Maniac* (1902), who having dealt with caddish motor dealers, finally had 'the intense gratification of making the acquaintance of two world-famed automobilists', 'Pellin Sedge' and 'Charlemagne Parrott' who helped her on her motoring journey.[99] Kennard's readers knew she meant Selwyn Edge and Charles Jarrott, both household names.

However, such narratives tend to underplay the wider role played by women as ambassadors and entrepreneurs for the sports of cycling and motoring. Clarsen has described the professional taxi driver Sheila O'Neil, practising in London around 1908, and the difficulties she had with the authorities as a woman trying to carry out her trade.[100] Mary Kennard herself had been a motorist since about 1899 with her chauffeur Brooks on a 3hp 'Ideal' Benz. She went on to drive models by now long-gone brands such as Napier, De Dion and Progress.[101] She was one of some dozens of female motorists, usually society women or actresses, identified in 'portraits' in the higher-class motoring magazines such as *The Car Illustrated* and *Motoring Illustrated*.[102] These 'motoristes' had a particular role in encouraging wider female participation in the mobility movements. Edge's first wife Eleanor (who was also a subject of the 'portraits') had been active as a cyclist in the Anerley Bicycle Club since at least the 1890s and played her part in the promotion of the Gladiator motor-car brand (provided by one of Edge's agencies, of course) in flower fairs.[103]

Female writers contributed to the motoring and society press too, including Mary Kennard. They tended to concentrate on matters of dress, etiquette and maintaining femininity – Mary Jeune's contribution to Harmsworth's Badminton book is a case in point – but women on the road, particularly by themselves, had for some time presented huge social challenges which needed to be addressed. Women had been riding bicycles from the outset (late 1860s) but, as Ritchie points out, usually in a risqué, circus-type environment. With the arrival of the tricycle or 'sociable' (that is, cycles intended to carry two people), women appeared, often with a partner, and usually doing so as a leisure activity, while the 'safety' bicycle brought respectability to the solo female middle-class rider by the mid-1890s. However, this use was bound by a series of conventions, notably the restriction on what they wore and where they went – the solo female cyclist, and the wearing of 'rational' dress, continued to offend sensibilities even into the twentieth century. There were very few competitive women cyclists – one exception being Helene Dutrieux (1877–1961) in the late 1890s, who

---

[99] Kennard, *Motor Maniac*, pp. 137–9.
[100] Clarsen, *Eat My Dust*, pp. 30–2.
[101] *Motoring Illustrated*, 22 March 1902, p. 64.
[102] *The Car Illustrated*, in particular, featured actresses and society women on its covers, for example, the Countess of Warwick, 11 June 1902.
[103] Fund-raising competitions were held where motor vehicles were draped with flowers and bunting. On another occasion his wife appears in a decorated Napier (supplied by him) in the 'Battle of the Flowers' at Earls Court in 1902: Duncan, *The World On Wheels*, ii, p. 840.

went on to race motor-cycles and motor cars, and became an early aviator.[104] Female racing drivers were also thin on the ground: Helene van Zuylen (1863–1947) was the first female driver in a motoring competition (the Paris-Amsterdam-Paris of 1898), and as with Camille du Gast (1868–1942), was skilled and able but from a privileged background. More difficult to socially locate was Dorothy Levitt (1882–1922), who encouraged a wider female participation through her newspaper columns (*Daily Sketch*) and her *Woman and her car: A chatty little book for women who motor or want to motor* (1909). Coming to motor racing and long-distance driving from a sporting background of horse-riding and cycling, Levitt nurtured her image of 'the most girlish of womanly women'[105] with stoicism and clear ability.

There were plenty of female columnists, notably Mrs C. C. Cooke, who contributed to *Motorcycling* magazine and was herself an accomplished rider and trialist. In one key article in 1910, she identified machines suitable for women and a dozen or so 'well known lady motorcyclists' active in trials or regular motor-cycling – Muriel Hind (1882–1956), middle class and sporty, is one name to recur.[106] 'Women's specials' of magazines were published to encourage the notion of the motoring female, and features were made of women such as Alice Neville, who had her own 'garaging and hire' business in Worthing – 'she is a capable driver, and is evidently a competent business woman'.[107]

Political inclination

Edge made no bones about his political leanings and his attitudes to labour and labour relations were probably typical of his peers. He was a Conservative Unionist, a protectionist, and scornful of trade unions.[108] He would probably have approved of a cartoon in *Punch* in 1902 which attached some blame for the failure of the country's motor industry to compete on the trade unions. In an image of a sleeping car, emblematic of British enterprise, it was 'fitted with a powerful trade union brake'. To labour the point, the car had steam-roller wheels too.[109]

Edge was clear in associating his entrepreneurialism with a particular flavour of politics, and it is likely his political values were formed in his cycling club days; here, Fry, president of the ABC in the 1890s, was a vocal Conservative. Indeed, Edge also

---

[104] Ritchie, *Early Bicycles and the Quest for Speed*, p. 365; Thomas Ameye, Bieke Gils and Pascal Delheye, 'Daredevils and early birds: Belgian pioneers in automobile racing and aerial sports during the Belle Epoque', *International Journal of the History of Sport*, 28:2 (2011), pp. 205–39.
[105] C. Byng-Hall, 'Dorothy Levitt: A personal sketch', in Levitt, *The Woman and Her Car*, p. 4. For du Gast and Levitt, see the individual chapters in S. C. H. Davis, *Atalanta: Women as Racing Drivers* (London: G.T. Foulis & Co, n.d. [c. 1950]).
[106] https://www.gracesguide.co.uk/Agnes_Muriel_Hind, accessed 21 April 2020; see for example her 'The fashion of the future', *Motorcycling*, 18 October 1910, pp. 578–80.
[107] *The Motor*, 2 September 1913, n.p. Clarsen also briefly describes Neville: Clarsen, *Eat My Dust*, p. 32.
[108] Edge was also a signatory to the British Covenant for Ulster, allied with the Primrose League/Women's Unionist and Tariff Reform Association: *Daily Mail*, 27 January 1910.
[109] *Punch*, 25 June 1902 (coronation number), p. 461.

shared Fry's healthy eating habits, and may have seen Fry as a role model or icon.[110] Politicians at that time tended to be in one of two camps: pro-free trade or protectionist. The general election campaign of 1906 hinged on this, and led to a Liberal landslide in part because the Liberal message was that protectionism would raise the price of bread. Each newspaper had its political bias, and the association of Edge with the Conservative *Daily Mail* was a natural one, particularly since it was owned by pioneering motorist and Conservative-supporting Alfred Harmsworth (later Lord Northcliffe). Two other senior Conservatives feature in Edge's life: the fifth Earl Howe (1884–1964), a racing driver in the 1930s, wrote the foreword to Edge's autobiography, which Edge dedicated to the then late politician Earl Balfour, an early motorist, cyclist, and customer of Napier cars. Both Howe and Balfour were Conservative MPs (indeed, Balfour had been prime minister). Edge's early connections with business and Conservatism is apparent: his one-time patron Harvey du Cros and Du Cros's son Arthur (1871–1955) were also Conservative MPs. Ernest Instone had, like Edge, served terms as president and vice-president of the SMMT; he was a Unionist.[111] Meanwhile, at the time of the general election in December 1905, an advert for Napier cars declared them 'British and suitable for elections'[112] – Conservative canvassing, no doubt.

'Tariff reform' (or, protectionism) was of such moment that it was discussed in *The Engineer* and the motoring press too. Edge gave public talks supporting tariff reform – in 1910 he spoke at Nottingham – and contributed £1,000 to the cause.[113] If there was tariff reform, he said, a foreign company he knew would erect a factory in this country and employ a thousand 'English workers'.[114] His letter to the *Daily Mail* in 1910 (at the time of the general election) said tariff reform would create more jobs, while 'Free Trade and a Socialistic government' would bring fewer of them.[115] Edge was always good for anti-Liberal comment in the *Daily Mail* and elsewhere. He was scathing of the Liberal government's innovation of labour exchanges, arguing that there were too few jobs and too many unemployed, and that was because of free trade.[116] *The Times* dedicated a eulogy to him in 1907, admiring his protectionist stance.[117] His fear of any lurch to the left informed his political pronouncements. By 1915 he was warning in the *Daily Mail* of what must happen once the hostilities were over. Under the heading, 'The war after the war: Defence of our trade' he offered his advice on industry in general and urged that committees of businessmen be formed to protect existing trades and get those affected by the war back into business. He anticipated 'our workpeople' will be 'below par' and 'suffering from' large wages. He believed soldiers would find it difficult to adjust to monotonous work and the country could expect trade attacks from the United States and Germany. 'Each trade will have to lay before parliament the conditions that will enable it to live and grow big and strong again,' he said. 'If we are to

---

[110] Thomas, 'The leviathan of the turf'.
[111] Duncan, *The World On Wheels*, ii, p. 662; E. K. S. Rae, 'Unconventional portrait of the week: Mr Ernest Instone', *Automotor Journal*, 25 December 1909, pp. 1541–3.
[112] *London Standard*, 4 January 1906.
[113] Smith, *Selwyn Francis Edge*, p. 50.
[114] *The Times*, 13 January 1910.
[115] *Daily Mail*, 27 January 1910.
[116] *Daily Mail*, 11 February 1910.
[117] Eulogy in 'Motor matters', *The Times*, 9 July 1907.

remain the dumping-ground of the world after the war, we shall see times of industrial trouble and workless people that must lead to emigration on a greater scale than has been ever seen here.'[118] If Edge's views before the First World War were out of kilter with the free-trade Liberal government's, then they chimed better after the War when his 'trade attacks' threatened the economy. Echoing Edge, Wyatt wrote in the 1920s that a policy of free trade had allowed untrammelled sales of foreign motors, which compared badly with the United States whose policy of protectionism had led to its very large motor industry.[119]

Edge was also comfortable playing the nationalist card. He arranged, for example in 1908, for a 'puff' piece to go in the *Flag*, the book of the Union Jack Club published by the *Daily Mail*, ensuring readers knew the proceeds would go towards beds for soldiers and sailors. Edge also made sure a Napier was at the disposal of the editor of the *Daily Mail*,[120] and Napier Cars were advertised within. In 1908 Edge, as one of many in the Union Jack Industries League, exhorted people to buy British because of the unemployment. In common with other car makers, Edge identified prominent Napier-driving Britons in its advertising.[121]

Concluding remarks

S. F. Edge has been used here predominantly to illustrate the likely background and experience of the motor-traction entrepreneur. Edge had gone on to achieve an elevated status as a national hero through his international motor-racing wins, but otherwise was not so different from many of his peers. However, it is clear that there was no stereotype. For example, those selling motor cars were from all backgrounds. The socially lowly might call themselves 'salesmen', and need to go cold-calling, whereas the aristocratic were well placed to use their club membership to facilitate deals. The formative experience of the entrepreneur might have included a commercial education, or immersion in a commercial household, as was the case with Edge. But most likely it was an experience of the cycling club – note, for example, how the motoring magazines had spotted so many of 'their' people as coming from the cycling world. The cycling club provided the means to compete at a national and international level, and connections made here gave a thorough grounding for subsequent dealings and behaviour. Edge set several records while in the Anerley Bicycle Club and in turn endowed the club with cups and served as its president.

Another likely trait was the willingness to build up knowledge by trial and error. This meant that on the earliest cycling and motoring trips, difficulties and roadside repairs were to be expected, obliging hard and filthy work, and often with a hostile public looking on. The knowledge gained brought the ease and efficiency with which the entrepreneur appeared to tackle all calamities. This, though, was part of an image

---

[118] *Daily Mail*, 1 December 1915.
[119] Wyatt, *Common Commodities and Industries*, pp. 1–4.
[120] *Daily Mail*, 15 May 1908.
[121] *The Times*, 16 November 1905.

often cultivated, for example the playing up to a role befitting the masculine hero of the time. This might have meant affecting nonchalance – Edge was very good at this – and building up an image through stories of fantastic feats, having glitzy business premises, and being seen in smart and fashionable suits with the socially elite. Edge also saw advantage in affecting a public accessibility and sympathy, for example, his voluntary marshalling at trials and association with charities for the poor.

Entrepreneurs were mostly male, but there was a great deal of engagement and initiative by women, such as authors, journalists, actors, socialites and businesswomen. However, helping pave their way were the countless, and often nameless, women who rode bicycles, sometimes wearing 'rational' dress, in the face of public derision. Their visibility helped encourage changing attitudes towards the wider mobility of women. Finally, a brief examination of Edge's political leanings suggests a clear inclination for Conservatism, wanting the state to enact protectionist policies to help build up a domestic motor industry, and with an emasculated trade-union representation (or other 'socialistic' influences, as Edge called them) to help allow this to happen.

The entrepreneur faced the difficulty of selling a product for which there were no brands, no infrastructure, no industry and practically no demand. To help counter this, they organized public trials to demonstrate the abilities and potential of motor traction, as seen in the following chapter.

4

Trials

The appearance by Dorothy Levitt as the only female competitor in the Automobile Club's Light Car Trials in August and September 1904 caused something of a stir. Her fame as a driver, but particularly a female one, in speed and long-distance motoring trials no doubt preceded her, and her association with the up-market Napier brand and with S. F. Edge, by then a national hero as the winner of the Gordon Bennett Cup, would have added to the allure. The six-day trials in which she was competing were based in Hereford and drew large crowds, as photographs attest. She brought along her black Pomeranian for the ride on her De Dion, so, as a joke, on the Thursday of the trial other competitors bought toy dogs to lash to their bonnets. The humour extended into that evening when the male competitors attended a smoking concert, at the interval of which Levitt had arranged for packets of dog biscuits, 'for the refreshment of the puppies', to be distributed with her compliments. Levitt's gesture led to her being invited to join the men in the concert as an 'honorary gentleman', where the 'K9' society was formed.

For these Light Car Trials, only motor cars costing up to £200 could be entered, so this was truly a test for the lightest and cheapest cars then available. There were four classes, divided by price, and thirty-eight vehicles entered, of which three were non-starters. Each day a circular trip of about one hundred miles was navigated, and the results assiduously collated. Out on one of these runs, Levitt was presented with a bouquet of flowers; she had inevitably drawn attention to herself out in the sticks too. A second De Dion had been entered into the Trials, and on the conclusion of the event a silver medal was awarded to De Dion 'for excellence of workmanship, for hill climbing and consistent running'.

The event concluded with a procession of the motor cars through the town, albeit without Levitt, whose car had developed a mechanical problem. All the cars displayed flags and flowers –– a deliberate gesture to help associate motor traction with a wider spirit of patriotism and empire. The trial was, by any measure, a success. The Automobile Club went away and totted up its results for publication, entrepreneurs and agents were now in a position to trumpet the appearance in the trial of their products, and manufacturers were able to improve their models in the light of any failures (Vauxhall, for example, learnt not to make their connecting rods out of phosphor bronze). The good people of Hereford had had a thorough opportunity to observe motor cars in action, speak to the drivers and draw their own opinions on motor traction. Finally, decorum had been resumed at the smoking concert on the final evening, to which only men were invited.[1]

In the United Kingdom, trials had become an established means to test motor cars and motor-cycles in the public arena. Many were held each year under the auspices of the Automobile Club, or one of its affiliated local clubs, and were intended to test vehicles on an increasing array of qualities. This was initially the fundamentals: reliability and durability, hill-climbing, re-starting on hills and effectiveness of brakes. As these abilities became more certain, testing began to include measuring the kicking up of dust in the vehicle's wake (flour was spread on a track, and the vehicle driven over it to assess how much dust the vehicle would raise); fuel consumption, smoke and noise. Usually conducted on the public highway, members of the public could look on, while journalists and photographers recorded the event. As seen in Hereford, celebrities made appearances, sometimes as drivers, or as onlookers. The reporting on trials was a mainstay of the popular magazines such as *The Motor* and *The Autocar*. Manufacturers, agents and magazines liked them because they offered the potential for exposure of a product, for brand-creation, magazine copy, triumphant advertising copy, and a means for developing an otherwise tender product. (The risk for the agent, of course, was that the vehicles might be rather publicly exposed as unreliable.)

This chapter assesses the place of trials as a forum for entrepreneurs and customers to have their say. The Thousand Mile Trial of 1900 was the Automobile Club's chance to bring the latest exemplars of motor traction to the public. In the years following, and once the public was more accustomed to the motor vehicle, whose basic flaws were soon fixed, the problem then was keeping trials relevant and useful. Consumers, though, played their part by identifying their wants, needs and gripes, and for the latter, it is shown that customer dissatisfaction featured highly. A case study is made of *The Motor*'s aborted attempt in 1903 to stage its own trial, a venture that underlined the supremacy of the Automobile Club as the national organizing body, but also exposed its increasing detachment from the attention of the non-clubby motorist.

The Thousand Mile Trial: Providing information for press and consumer

The Thousand Mile Trial of April and May 1900 was an opportunity for the Automobile Club to take the motor car to the public. On a scale never before attempted, it had been first suggested in 1899. A circular route was mapped out which started and ended in London, via Edinburgh. Each day on the road covered about one hundred miles, sometimes with optional hill-climb tests. The route was reconnoitred in mid-winter by motor car by Claude Johnson and Montague Grahame-White, itself a huge physical

---

[1] Gordon Brooks, 'The Small Car Trials of 1904', *Aspects of Motoring History*, 3 (2007), pp. 18–39; Donald Cowbourne, *British Trial Drivers: Their Cars and Awards, 1902–1914* (Otley: Westbury Publishing, 2003), pp. 251–73. For a useful analysis of trials, see T. R. Nicholson, *Sprint: Speed Hillclimbs and Speed Trials in Britain, 1899–1925* (Newton Abbot: David and Charles, 1969), esp. part 2.

Figure 4.1 The Thousand Mile Trial of 1900 stopped off in Manchester, permitting respectable-looking members of the public to inspect the vehicles. The photograph was taken by Argent Archer. Image: Malcolm Jeal.

and mechanical challenge.[2] Trials were initially intended to show the viability of the motor car: and here, in full public glare, to show how the motor car could do such big distances, way beyond the possibility of horse traction. It also brought the motor car to places where one had never been seen before.

Unlike previous Automobile Club events, the trial was underwritten.[3] Previous attempts by the Club 'to interest the public in motorcars', such as the Richmond Show in 1899, had been described by Claude Johnson as a 'colossal failure'. Richmond lost the Club £1,600 and showed, three years on from Lawson's 1896 'Emancipation Run', that 'the public was [still] not at all impressed with the merits of the new mode of locomotion'. Then, Johnson said,

> the motor car was looked at askance by all important people. It was regarded as only suitable for those who minded not dirt and discomfort – something for the lower orders to play with (within restricted bounds) and for the upper classes to sneer at, and upon which the governing section could exercise its repressive influence.[4]

---

[2] See St John Nixon, *Romance amongst Cars* (London: G.T. Foulis, 1937), pp. 131–8; Montague Grahame-White, *At the Wheel Ashore and Afloat* (London: G.T. Foulis, n.d. [*c.* 1935]), chap. VI; *The Autocar*, 6 January 1900, pp. 22, 44–5.
[3] *The Motor*, 19 April 1910, pp. 422–7, 445; 26 April 1910, pp. 471–3.
[4] *The Motor*, 26 April 1910, pp. 471–3.

Edge recognized that motor cars at that time were a 'novelty', with public faith 'at a low ebb'. He pointed out how road conditions remained dreadful, with the dust raised being so bad that he had to rinse his eyes at night. Tyres were awful, with no standard at that time even for the design of a valve. In addition, the spare battery was often flat, the chain ran in the mud, the use of candle lamps was normal with acetylene seen as futuristic, and mudguards and cooling systems were rudimentary.[5]

The public also needed to be reassured that the trial would not turn into a race; rumours persisted that the vehicles returning from Brighton in the 1896 'Emancipation Run' had had a 'motor race', reaching as much as 30mph downhill.[6] Something sensible like a trial, reasoned the *Motor Car Journal*, was 'most likely to appeal to the practical character of the British people', because there would be no 'racing at fearful speeds, on high roads' as they did in France and Belgium.[7] As a result, a barrage of rules and regulations, intended to 'advance the Automobile movement in the United Kingdom', were put in place by its principal organizer Claude Johnson, as detailed by Bennett.[8] These were meant to 'prevent the passage of the Trial vehicles being a source of annoyance or danger to other users of the road', to which end they 'earnestly begged' the drivers not to exceed the speed limit. (This did not prevent Charles Rolls, in particular, from seeking to be first on every day and in every event.) There were opportunities at Bristol, Birmingham, Manchester, Kendal, Carlisle, Edinburgh, Newcastle, Leeds, Sheffield and Nottingham for paying members of the public to see the vehicles close-up, at exhibitions which also included sideshows such as stalls by accessory agents, plus 'fancy' driving displays, such as up and down steep ramps, to demonstrate the abilities of the motor car.

The vehicles entered ranged from motor-tricycles of 2¼hp to powerful 12hp motor cars, in either 'trade' or 'private' (amateur) class. For the trade entries, thirty companies entered forty-five vehicles, of which thirty-two finished; for private, twenty-two out of twenty-four entrants finished. The official photographer was Albert Argent Archer (1860–1932),[9] who also covered the Hereford trial in 1904. He accompanied the trial, along with journalists who were offered seats in participating cars. A gaggle of cyclists followed the route, including F. T. Bidlake (1867–1933), a one-time racing cyclist. A range of prizes were offered, with twenty-five *Daily Mail* prizes of ten pounds plus a range of gold, silver and bronze medals, cups and clocks for each class (which was determined by the sale price of the vehicle).[10]

Automotor Journal drew attention to a clear public interest. It reported how, in the cities and towns, the footpaths and roads were so densely crowded with spectators that only a narrow passage remained. At every crossroad in the country, there were knots of onlookers from neighbouring villages: 'the parson and his daughters on bicycles, the country squire on his horse, the old dowager safely ensconsed in her landau, coaching parties enjoying champagne lunches at the roadside, and cyclists in legions'. In villages,

---

[5] Edge, *My Motoring Reminiscences*, pp. 78–80.
[6] Letter to *The Times*, 20 November 1896; 'Motor-car speed', *Daily Mail*, 20 November 1896.
[7] *Motor Car Journal*, 27 April 1900, p. 131.
[8] See Bennett, *Thousand Mile Trial*.
[9] https://www.gracesguide.co.uk/Argent_Archer, accessed 13 May 2020.
[10] Bennett, *Thousand Mile Trial*, passim.

the children were given a 'whole holiday' and were ranged on the school walls, cheering each vehicle as it passed. Crowds assembled at the bottom of hills leaving 'barely 7 feet wide through which the vehicles had to pass at high speed'. 'Generally, the public looked on the passage of motors as they do on the passage of a fire-engine, namely, as a fine inspiring sight which makes the pulse beat faster and satisfies a craving for excitement.'[11]

Yet the reception as the caravan passed through the English and Scottish towns and countryside was mixed. In its 'Motors and Moting' column, *The Clarion* commented on the motor cars passing through Buxton, where most of its 'beauty and chivalry' had turned out to watch. Of the motorists, some 'were garbed in the most outlandish rig. There were men clad cap-a-pie in leather of proof, with rococo ornaments of limestone dust and oil. Many wore unsightly goggles of coloured glass, and the costumes of the ladies looked more comfortable than elegant'. The correspondent continued, 'Most of the cars emitted a not altogether inodorous reek', with the steam-driven ones worst, 'one enveloped in a dense cloud of sulphurous smoke and profane opinions'.[12]

The muddy roads created a problem of image. On their first day,

> one feature of the gathering could not escape notice. Everyone was grimed with indescribable dirt. What the cynical section of the public must think of the apparitions whizzing past them in whirlwinds of dust is probably unsuitable for publication, but any motorist of half an hour's experience knows at once that motoring means ruin to one's appearance.[13]

When passing through Darlington, what was intended to be a 10.30-prompt 'triumphal march through the town' was, instead, a string of cars and tricycles, arriving late, 'mud-besmeared, and owing to the wretched weather presenting altogether a sad spectacle'.[14] *The Autocar* had a ready explanation, regretting that an association has been made between the Thousand Mile Trial and dirty driving, ergo, motoring is dirty. It must be confessed, it said, they are arriving 'as grimy as powder monkeys', but that was because they were in close company, with hundreds of bicycles, and 'a sort of sirocco' was whipped up.[15] For a while, motor clubs had attempted to cast motor traction as impervious to the filth of the road (and, therefore, somehow, better than the horse). *The Autocar* reported that it had been suggested participants have handbills to distribute as they entered towns to account for why the cars were so dirty. Harmsworth subsequently offered a five-pound cup for the cleanest car to arrive at Southsea Common during the Automobile Club run from London in November that year. 'There is no question that the general public are not impressed by the advent of a motor car into their town covered with dust or mud. It never strikes them that the

---

[11] *Automotor and Horseless Vehicle Journal*, 27 April 1900, pp. 411–12.
[12] *Clarion*, 5 May 1900, which inaccurately said some of vehicles were electric.
[13] *Bicycling News and Motor Car Chronicle*, 2 May 1900, p. 20.
[14] *Motor Car Journal*, 12 May 1900, pp. 174.
[15] *The Autocar*, 12 May 1900.

Figure 4.2 At selected towns en route, the vehicles of the Thousand Mile Trial were put on display to a paying public. Here, the vehicles are on display in the Royal Botanical Gardens at Old Trafford in Manchester, 1900. The photograph was taken by Argent Archer. Image: Malcolm Jeal.

carriage [motor car], has, perhaps done its hundred miles.' The smarter a car looks, 'the better it is for the furtherance of the automobile in this country'.[16]

An association of elegance with cleanliness prevailed, which was very difficult for motor-car owners to sustain. Bidlake put his finger on it: 'how filthy you all are, folk say. Smothered with dirt, and quite different looking from carriage folk.'[17] *Hearth and Home* magazine found the appearance of the cars as they arrived back in London after the Thousand Mile Trial 'decidedly amusing'. 'For the sake of the speedy advance of motor car touring', it continued, 'we could have wished that those engaged in the thousand miles trial had looked a little less like the members of a 5th of November procession'.[18] The Automobile Club had laid down rules on how much time could be spent maintaining the vehicles at each overnight stop, so that the public might be persuaded they really did not need much maintenance, but this meant limited opportunities for the participants to clean their vehicles and do essential repairs.

Once the Thousand Mile Trial finished in London, there were further opportunities for agents and manufacturers to place their motor cars in public view for a week at the Crystal Palace. Overall, the press coverage had been more measured and generally positive than for the 'Emancipation Run' of four years earlier. The writer for *Society*

[16] *The Autocar*, 3 November 1900.
[17] *Bicycling News and Motor Car Chronicle*, 2 May 1900, p. 21.
[18] *Hearth and Home*, 24 May 1900, p. 146. See Horner, 'The emergence of automobility in the United Kingdom', pp. 56–75, esp. p. 67.

noted how inordinately anxious the Automobile Club seemed to be to prove that the motor-car 'can do a journey without blowing up or coming to pieces'. But it reckoned that 'motorism' was 'the inevitable mode of travel, and perhaps, of transport, in the not too distant future', and that 'we none of us need convincing of the merits', but observed it was 'our ancient conservatism, and perhaps our natural love of horseflesh, [which] restrains us from committing ourselves to "horseless traction"'.[19] *Automotor Journal* noted the excellent organization, saying the Trial had 'attracted widespread attention from the public, and that it has met with hearty approval from them, is equally certain'. It was sure many benefits would follow, one being the 'effect produced upon the mind of the general public', but more optimistically that it would 'demonstrate the fact that motor cars have ceased to be the dangerous, noisy, trembling and smelly affairs' of their imaginations.[20] *The Times* presented a paternal viewpoint more in tune with the views of the privileged few then motoring. 'The principal object of the trial was to prove', it said,

> what it was considered the people of this country needed to be taught – that the motor car is, even in its present state of development, a serious and trustworthy means of locomotion . . . destined to take its place with the train and bicycle as a common object of daily life, and as superior to them, in many respects, as they are superior to the horse and cart.

The same newspaper also observed that owners of horses had 'sent them to stand in their rugs by the wayside and "get accustomed to the things [motor cars]"', and in a master-stroke of patriotism, the Automobile Club ensured that proceeds from all of the public showings – some several hundred pounds – went to the Transvaal War Fund.[21]

The Trial presented opportunities, usually through motoring magazines, for agents to advertise, inform and sway public opinion. The 3½hp Triumph, entered by the Motor Car Co. of London was a case in point. This company acted as an agency for motor manufacturers and was the initiative of the Farman brothers, one-time amateur racing cyclists, now with connections in Paris and later involved in aviation.[22] Their advertising of the Triumph following the Thousand Mile Trial did not hold back on hyperbole:

> SENSATION OF THE TOUR: You noticed how it climbed the hills in these valuable competitions. Your Autocar [the magazine] had it down in black and white. Fancy climbing hills of 1 in 15 at 10 miles an hour. If it didn't go up the hills what would be the use of it? FOR WHY IS A MOTOR CAR WHEN IT WON'T CLIMB HILLS? [. . .] What you can be quite sure of is that you can face any hill with absolute confidence, and if the LAW'S EYES are not upon you, tear along at 30 miles in the hour. SHALL WE SEND YOU FURTHER PARTICULARS?[23]

---

[19] Cited in *Motor Car Journal*, 12 May 1900, pp. 169–70.
[20] *Automotor and Horseless Vehicle Journal*, May 1900, p. 408.
[21] 'The 1000-Mile trial of motor cars', *The Times*, 14 May 1900.
[22] Bennett, *Thousand Mile Trial*, p. 73; https://gracesguide.co.uk/Henry_Farman, accessed 27 September 2019.
[23] *The Autocar*, 19 May 1900, ads, p. 3.

As Brendon has shown in his study of the Automobile Club, trials were part of an ongoing public-relations exercise which legitimized the national club[24] and the local clubs which had assisted it; the Herefordshire Automobile Club had been the driving force behind the Hereford event. Trials also enabled fault-finding and incremental improvements for the manufacturer, while providing exclusive social outings for the participants, as well as a sense of adventure and competition. Magazines sent reporters to file accounts of the meetings of some of the bigger clubs (such as the Manchester Automobile Club), and those reporters occasionally secured a seat for the day on a car driven by a celebrity such as Edge; for such a privilege, copy could therefore be expected to be tame. Photographers, most famously Argent Archer for the Thousand Mile Trial in 1900 – and for the Light Car Trial of 1904 – were commissioned to record the proceedings. The opportunities for meeting fellow enthusiasts, for checking out the competition, for parading the latest models, all presented themselves.

Agents and manufacturers used trials as an opportunity to create a brand even when the results were indifferent. The International Motor Car Co. of London had entered two 3hp Victoria motor cars and placed a notice in *The Autocar* while the Thousand Mile Trial was underway in apparent expectation of both a dismal performance and public scepticism:

> TWO Internationals are taking part in the little jaunt of the 1000 Miles trial run. If our two cars in this competition were so fortunate as to beat every other car of the eighty odd entering, IT WOULD PROVE VERY LITTLE. Because one car of any particular make will do certain feats, it is absolutely no guarantee that others of the same brand will do the same. COMPETITIONS ARE USEFUL in one sense. They prove that the motor car is a really serviceable and practical article; that they can be bought with every reason to believe that they will do the work they are intended to do; that the motor is not 'in its infancy', as one so often hears the wiseacre remark. They are not 'the coming thing, no doubt.' 'They are the COME thing', and what's more, they have COME TO STAY. BUT REMEMBER, whether Smith's, Brown's, Jones's, or Robinson's car does best in this competition, it does not in any way prove their motor to be the best. Many cars might be got up specially for this competition, consequently would probably be superior to the ordinary stock car'.

It continued rather more confidently: 'EVERY STOCK INTERNATIONAL is guaranteed to be capable of doing anything and everything which is required to be done in this competition, and the two cars used are guaranteed to be absolutely stock patterns cars, and in no way specially made for the occasion'.[25]

As it happened, both International Victorias finished (although with some under-average speeds on several days), meaning in the following week they were a little more braggish: 'and the most serious thing that happened was the loss of two lubricators,

---

[24] Brendon, *The Motoring Century*, esp. chap. 2.
[25] *The Autocar*, 12 May 1900, ads, p. 7.

which is a mere trifle, and not worth considering. Even this would not have happened had the driver of the car (No. 41) been careful to screw them in tightly in the morning'.[26] Meanwhile, advertising for the MCC Triumph trumpeted: 'its journey from London [...] to Leeds (at the time of writing) has been one of conspicuous success';[27] yet, of the two MCC Triumphs entered, one came second in Class B; the other finished but had failed to register for several sections. The advert for the Orient Express declared it 'First again' in its section (Cheltenham to Birmingham), 'beating cars three times its price'. Yet, in this instance, the car had failed to finish the Trial following a collision with a cow on the Carlisle to Edinburgh section of 2 May.[28] And then Star were able to make a virtue out of a disadvantage: 'THE STAR MOTOR CAR will climb any hill: 9⅜ miles an hour up Taddington, and 8 miles an hour up Birkhill in the Hill Competition. THIS WITH ONLY TWO SPEEDS.'[29]

The intending purchaser present at a trial could see the vehicles being put to the same test in the same conditions, and in due course read the results in full in the magazines, and draw his or her conclusions. Anomalies did not go unnoticed. A letter from 'Fairplay' in *The Autocar* asked why E. H. Coles on his Benz got the first prize in the Thousand Mile Trial. He had had several serious breakages, including the breaking of the front axle, and having a new one fitted; the main bridge of the bar breaking, which supports the motor, having to be fitted with a plate to support it; the main frame breaking at the near end; and the cracking of the main arm of the motor supporting the flywheel bearings. Coles had 'exceptional aptitude for "quick repairs"', but for the 'ordinary man', they would have rendered the car 'hors de combat'. Surely, 'Fairplay' continued, these breakages should disqualify the car, when other cars did nearly as well, speedwise, with no breakages? 'I really cannot see the practical use of such a competition to the public'. The second prize, the Locomobile, failed to make a report on one day and failed to make the full speed on five other days.[30] 'A. D.' found the judges' report 'unsatisfactory':

> As an outsider, knowing little or nothing of motor cars, I had patiently awaited its publication in the full expectation that it would have guided one in the selection of a suitable one. The report withholds the reasons why those cars failed to report themselves, or to keep to the time limit. This is the 'very pith and essence of the whole thing', because if a car fails because of its machinery, or needs parts, then that is different to having a puncture.[31]

The Automobile Club had prepared well for this trial, but it had proved difficult to log every event and time, and then weigh up the findings afterwards. However, *Motor Car Journal* magazine believed that the Thousand Mile Trial of 1900 was truly a proper

---

[26] *The Autocar*, 19 May 1900, ads, p. 7.
[27] *The Autocar*, 12 May 1900, ads, p. 3.
[28] *The Autocar*, 12 May 1900, ads, p. 2.
[29] *The Autocar*, 12 May 1900, ads, p. 6.
[30] *The Autocar*, 14 July 1900, pp. 678–9.
[31] *The Autocar*, 28 July 1900, p. 725.

endurance test, aware that even on stationary days, drivers were allowed only limited time for cleaning and maintenance. It decided that

> The beneficial effects of the trial are beyond measure. To the public it has revealed the fact that motor-cars are by no means toys, but are vehicles which can be relied upon to do their work efficiently; to the manufacturer it has revealed weak points which needed improvement and modification in a way that no isolated trips of 100 miles or so could possibly do. It has been said that many would-be purchasers of motor-cars have deferred making the plunge until the Trial was completed.[32]

The Automobile Club, in part through the agitations of its members with commercial interests, had been able to attract sufficient commercial interest in the Trial. It announced in *The Autocar* that 'trade readers' taking part in the Trial would get 'infinitely better advertisement than by taking part in any stationary exhibition', which had been the usual practice hitherto. Paying the fee to enter the Trial – £20 early entry, or £40 late entry, neither refundable in the event of a no-show[33] – also included free exhibition in seven major cities en route plus a week on display at the Crystal Palace afterwards. 'Although the machines may not look quite so spick and span as when run into a covered exhibition directly they leave the painter's hands, each car will bear absolute and positive testimony as to its practical value as a road machine, and after all, that is what the buying public are after',[34] the Club suggested. *The Autocar* had described all of the motor cars for which an interest had been expressed in taking part, including those which did not ultimately participate. For example, the official programme of the Trial included adverts for motor vehicles of which more than a dozen makes did not go on to start.[35] The Trial, then, was seen by manufacturers and agents as worth a punt, even if they did not end up participating, and so it proved.

Motor racing on the public highway had never been allowed in the United Kingdom, but was permitted in continental Europe. The races were often spectacular and glamorous events, drawing in the racing celebrities and best manufacturers. The 'town-to-town' races were particularly popular at the turn of the twentieth century, having started with the Paris-Bordeaux-Paris in 1895, and magazines ensured that teams of journalists and photographers covered them.[36] The resulting reports helped create and reinforce reputations and brands. Magazine sales spiked when reports of races such as the Gordon Bennett Cup were published.[37] However, the tenor changed following the Paris-Madrid race in 1903, which was aborted even before it had reached Spain, following six deaths amongst competitors, spectators and marshalls. Racing cars had been getting faster and more powerful, and the dusty roads often meant that a competitor in the wake of another car was engulfed in so much dust that he could only

---

[32] *Motor Car Journal*, 19 May 1900, p. 191, reflecting on the Thousand Mile Trial.
[33] Bennett, *Thousand Mile Trial*, p. 46.
[34] *The Autocar*, 3 February 1900, from Bennett, *Thousand Mile Trial*, p. 48.
[35] Bennett, *Thousand Mile Trial*, pp. 48–50, 56–8.
[36] Particularly useful here is Rose, *A Record of Motor Racing 1894–1908*.
[37] For example, circulation of *The Motor*'s 1903 Gordon Bennett special was 50,115, well up on its regular circulation of 36,072: *The Motor*, 19 July 1903, ads.

navigate by looking up to use the tree tops as a guide to the line of the road. This race was the most deadly of any so far, although, as Jeal has shown, newspapers played up the image of carnage and death; *Punch*, for example, published a full-page cartoon, 'The race of death'.[38] With such negative reaction to what were inevitably seen as reckless European events, the feeling at home was ambivalent about the safety of public races and helped create an environment in which the trial was eminently sensible and appropriate. A 'common sense' approach prevailed; Robert J. Smith, for example, the honorary secretary of the western section of the Scottish Automobile Club, argued in 1905 that reliability trials were 'a powerful agency in educating the public'. In addition, the publicity of the trial had led to 'hundreds' following the trial route, such that 'hotel-keepers say motoring custom has increased fourfold', with indirect benefits of the upkeep of the roads (ensuring road-works were finished and road surfaces improved in advance).[39] Local clubs continued to stage reliability and efficiency runs – for example, the Manchester Automobile Club's mix of hill-climbs and reliability runs up until 1914,[40] and good results from local competitions had sufficient significance for brands to advertise nationally – for example, Vauxhall winning a Manchester Automobile Club trial and announcing this in *The Autocar*.[41]

It has already been shown how Edge was notably active in trials and hill-climbs and associated his name with prestigious prizes. As with Levitt, his presence at trials as a celebrity added no little stardust, and magazines would give notice of the attendance of any 'names'. He continued to sponsor trials and hill-climbs even after he had withdrawn from motor racing in 1908. Edge believed trials offered exposure for British products (for him, Napier) and that they were vital because the customer continued to want to know which car went fastest, including in hill-climbing;[42] trials provided these answers. He ensured good results were incorporated into the extensive Napier publicity, to be found in national newspapers, motoring-magazines and society magazines.

Sustaining interest in trials

Over time, though, for trials to remain meaningful, the challenges had to become more onerous. Within a few years of the Thousand Mile Trial there was the 5,000 Miles Trial; and another stunt was driving from London to Edinburgh in top gear. Here, while only larger cars could perform such a feat – travelling up hill and down dale using top gear only – it was advantageous for customers to know that a car could be used

[38] *Punch*, 3 June 1903, p. 389. See, for example, Malcolm Jeal, 'Facts spoil a bloody tale: the 1903 Paris-Madrid Race', *Aspects of Motoring History*, 1 (2005), pp. 32–5; Michael Ulrich (ed. Thomas Ulrich), *Paris-Madrid: das grosste Rennen aller Zeitend* (Münster: Monsenstein and Vannerdat, 2015).
[39] See H. Massac Buist (ed.), *The Motor Year Book and Automobilist's Annual 1906* (London: Methuen, 1906). The chap. 'Words for the wise and witty' by various contributors included Robert J. Smith who argued why reliability trials were still useful (referring to the Scottish Trial of 1905): pp. 300–3.
[40] See John Archer, 'The Manchester Automobile Club Reliability Run, 1907–14', *Aspects of Motoring History*, 13 (2017), pp. 49–70.
[41] 'A remarkable feat', *The Autocar*, 18 May 1912, ads, reprod. in Archer, 'The Manchester Automobile Club Reliability Run', p. 64.
[42] *The Times*, 16 October 1908, ad.

with minimal gear-changing, as this was one of the most demanding facets of motor-driving. But while all these trials continued to provide fodder for brand advertising and magazine copy, they were increasingly challenged by the readers. When Arthur Cecil Edge (S. F. Edge's cousin, known as Cecil, 1879–1908) accomplished his observed trip from Brighton to Edinburgh in 1905, all in top gear, 'One who wants to know' wrote to *The Autocar* to ask, 'What is the point? And why hasn't Mr Edge published details of the fuel consumed?'[43] His was not the only letter making the same observations.

The ensuing letters to motoring magazines give a sense of the spread of opinion amongst consumers. In 1905 'J. S.' thought that only a few readers were sufficiently educated to understand the way the magazine tabulated its results; these were complicated and laborious. 'J. R. P. G.' thought trials had no value. He had bought a 12hp motor car (which was not identified) based on the results of the trials and found it unreliable and noisy. '10hp' wrote to say he was in a position to buy a motor car in January 1904, and, knowing nothing of the market, bought a copy of *The Autocar* and mugged up. He was confused by the advertising, which suggested each brand was the best, so he asked friends, each of whom advised him to buy different brands. He even engaged a chauffeur, believing he would need one, but was able to teach himself in two days how to drive and dispensed with the man's services.[44]

Furthermore, and to the irritation of readers, the Automobile Club took weeks after its trials to collate its findings and publish any results. Another factor was that actual results for speed trials or hill-climbs risked 'proving' that speeding had taken place, so magazines tended to publish the name of the winning driver and car with a time of zero, with adjusted times for successive finishers. For the Easter Automobile Club tour in 1902 to Cromer, there was a speed trial at Gunton Park. Edge won this, but the report in the *Daily Mail* said the results would not be divulged to the public at that time, and, instead, would be intimated at the forthcoming Whitsun meeting.[45] Similarly, the eliminating trials for the Gordon Bennett Cup in 1903 were held at Welbeck Abbey, the estate of the Duke of Portland.[46] Details of the competitors on the open road the following day were not revealed until just before midnight and at Banbury, with the candidates then sent to Dashwood Hill by West Wycombe, where the fastest time given was zero with the others having a minus figure. It was not unusual, then, for motor-sporting venues to be revealed only at the last minute, and to travel there under cover of darkness, in order to thwart any police interference.

From the agent or manufacturer's point of view, trials and racing were a mixed blessing, requiring much investment of time and money. Herbert Austin, then at the Wolseley Company, complained how (in this case for the trials at the 1900 Midland Cycle and Motor Show) speed was the 'main factor' in the judgement, with no account made for details which would make the cars a commercial success. He said that some of the cars arrived filthy, meaning mud would have collected on the working parts; this accounted for why the Mors overheated twice, yet it was a cold day. 'You say the

---

[43] *The Autocar*, 18 November 1905, p. 628.
[44] Letter from '10HP', *The Autocar*, 18 March 1905, pp. 390–1.
[45] *Daily Mail*, 31 March 1902.
[46] Notice in *Daily Mail*, 10 April 1903; Nicholson, *Sprint*, pp. 40–1.

Wolseley's belt was too short and narrow', he said, 'but it could have been adjusted by the driver, and it still managed to pull 12cwt and two passengers.' Austin thought that the Liverpool Self-Propelled Transport Association's trial of heavy vehicles was conducted on a much more broad and proper basis.[47]

As motor cars became more reliable and standardized, the value of trials to customers was increasingly questioned. When the new magazine *The Chauffeur* appeared, its first editorial promised to cut through the dross on the market:

> If it were not for big useless races, reliability, hill-climbing, and a dozen other sorts of trials, the results of which have to be worked out by professors of algebra, trigonometry and logarithms for a week before anybody knows what it is all about; if it were not for pretty pictures of Miss Maudie Fitztights in *her* motor car, of pretty castles and gardens belonging to some big swell who happens to have a motor car [...] where would most of them be?[48]

Always keen for copy which would interest their readers, editorials in the motoring press raised the issue of the value of trials. *The Autocar* in 1905 pointed out that by this time most manufacturers were now against trials, because of the expense of taking part, and the minimal returns from doing so, while three in four readers couldn't care anyway. The magazine did point out, though, that the participants picked up much knowledge by running the cars daily, a requisite of trials by then. And those which did well in the trial were usually the better-selling cars, possibly as a result of the reputation gained in previous events. *The Autocar* believed trials gave confidence to the would-be automobilist who would follow that class of car of interest to him. The magazine would be sorry, though, if trials were discontinued, although it imagined that subsequent purchases would probably be made without reference to trials, when the buyer had confidence in his own judgement and had taken the advice of friends.

Many readers, however, claimed not to be taken in by the results. *The Autocar* picked up on frequent bafflement among its readers when two identical cars were entered by the same manufacturer, yet did not finish with identical results. Just because a car wins the Gordon Bennett Cup, said 'J. T. H.' in the letters columns, does not mean that that company will make a reliable car, while 'P. H. H.' wondered how we could know the winners of the trials hadn't been 'faked'? An article in the same issue suggested it was 'well known' that the majority of manufacturers were against trials because they did not believe they influenced the automobilist in their choice. 'J. M. D.' even went as far as to say he would not buy a brand competing in a reliability trial. He said he had studied the details of other, non-competing cars, and preferred them. 'H. G. M. C.' said he regarded trials 'as [an] absolutely useless waste of time and money'. You can see that when twin cars are entered, one does well, the other doesn't finish.[49]

---

[47] *The Autocar*, 17 February 1900, pp. 166–7. For the Liverpool SPTA trials see Butt, 'The diffusion of the automobile in the North-West of England, 1896–1939', chap. 2.
[48] *The Chauffeur*, 9 November 1907, p. 1.
[49] *The Autocar*, 18 March 1905, pp. 383, 390–3; with further comment in 25 March 1905, pp. 442, 492.

Trials did continue, particularly organized by local clubs, but by the First World War they were out of fashion. Light motor cars, once considered unequal to the challenges of full trials, had caused *The Motor* to demand a different kind of test for them, with the result being the 1904 Light Car Trials. Such was the rate of product improvement since then that the editorial comment on the RAC Light Car Trial in 1914 recognized that despite dreadful conditions, this time over 1,010 miles in Yorkshire, nine cars came through perfectly.[50] *The Motor* in 1914 reported Arthur Stanley (1869–1947), the then chairman of the Automobile Club and a Conservative MP, saying that trials would now give no new lessons as far as the ordinary car was concerned, and that the recent light car trials had been 'very instructive'.[51]

The local clubs and public trials

Local clubs, affiliated to the Automobile Club, were highly active in staging their own trials, and the Manchester Automobile Club (MAC) and Manchester Motor Club (MMC) serve as case studies here. The MAC had assisted the national club to host the competitors in the Thousand Mile Trial of 1900 as they passed through Manchester. The caravan halted in the town for two nights, with a day-long public exhibition of the vehicles at the Royal Botanical Gardens at Old Trafford. Photographs taken by the official trial photographer Argent Archer show the vehicles arrayed in a glass-and-steel structure akin to the Crystal Palace. Visitors were admitted for a shilling on the open day, or could pay two shillings on the evening before and witness the vehicles arriving in the Gardens.[52] With an address by the lord mayor of Manchester, and dignitaries such as the chief constables of Manchester and Salford present, the event did much to raise the profile of the local club.

The MAC was, according to Archer, a 'gentleman's club for a privileged minority', with the ear of the local chief constables and politicians such as Winston Churchill (then a local MP) and Prince Francis of Teck (1870–1910), president of the Automobile Club. The MAC had been formed in 1899 and its 'leading lights' included solicitors T. W. Grace, a 'noted runner and cyclist'; and J. B. Thistlethwaite, an 'all-round sportsman'.[53] This, then, was the same sporting, clubby stock that had formed the mainstay of cycling clubs in the 1880s and 1890s (although a particular difference may have been the distinct Liberal persuasion of the MAC committee). The club initially had a familiar problem: many of its members were without motor cars. When *The Autocar* reported on the club's run to Nantwich in 1900, only four members turned out of its then fifty-two because 'some members have not yet purchased, whilst others have their vehicles on order, and are patiently waiting for delivery'.[54] However, reports only a few years later showed rather more activity; four runs by the club were reported in *Motorcycling*

[50] *The Motor*, 19 May 1914, p. 721.
[51] 'Car trials: Are they unnecessary?', *The Motor*, 2 June 1914, p. 811.
[52] See Bennett, *Thousand Mile Trial*, pp. 159–84; photos in the author's collection.
[53] Archer, 'The Manchester Automobile Club Reliability Run', pp. 49–50.
[54] *The Autocar*, 28 July 1900, pp. 715–6, report on the Manchester Automobile Club run to Nantwich.

and Motoring in 1902, usually heading into Derbyshire or Cheshire.[55] Soon the club was running all-weekend events and going as far afield as Dublin via the Holyhead ferry. By 1907, when it had 265 members, a captain had been appointed to take charge of social and competitive matters, and soon after, women were permitted to join.[56]

That same year the club began to stage reliability trials, which, according to Archer, became 'the premier motoring event of the North of England calendar'. These annual events, lasting from 1907 to 1914, covered a 132-mile route to Bettws-y-Coed in North Wales. From the outset, the club recognized the potential of these competitions for alienating the wider public, being seen to create problems on the highway with dust and nuisance. Intended to test for reliability, starting, petrol consumption and hill-climbing, the club emphasized that 'nothing in the nature of racing will be permitted'. As might be expected, there was a swathe of rules to follow. Vehicles were weighed, and the quantity of petrol in the tank measured at the start and finish. An official observer sat in each car to ensure the rules were observed. For the five stages, a minimum and maximum time was set, to prevent racing. For the 1907 event, only one car achieved the maximum of 1,000 points, and of the twenty-eight starters, twelve achieved 'non-stop' runs. The local newspapers and the motoring press picked up on the event, reproducing photographs and publishing the results. Manufacturers made much of any respectable performance in their advertising.[57]

The Manchester Motor Club, on the other hand, formed in 1900, was open to motorcyclists, motorists – and women – and had a more public profile with a much wider calendar of activities. This was much in line with its more 'democratic' character; it was, according to Archer, 'socially inferior' to the MAC. By 1905 it was staging events such as hill-climbs, reliability trials and speed judgement tests, all intended to raise the profile of motor traction and the club's popularity. The events were also categorized by trade and non-trade[58] in recognition of its commercial membership. The club's hill-climb in September of that year was held in Glossop, climbing Snake Hill, and had thirty-five entrants in five classes, with thirty-one finishers. For the hill-climbs, photographs in *The Motor* at Glossop showed an interested crowd, and the club's public activities would have attracted much attention. The club also took care to cultivate those people whom they wanted 'on side': the lord mayor and the chief constables, all of whom attended the MCC annual dinner that year (on the menu of which, to much amusement, was 'run-over chicken').[59]

As with the nationally run events, the public activities of the Manchester clubs had shown a keenness to portray motor traction in a positive light through the adoption of sensible trials and sporting events, to which the public and press would have come. The spectator was, for them, a potential recruit to the movement, and so it was vital that the events were seen to be conducted with professionalism and fair play.

---

[55] *Motorcycling and Motoring*, 11 June 1902, p. 301; 6 August 1902, p. 425; 17 September 1902, p. 93; 1 October 1902, p. 125.
[56] Manchester Automobile Club, *Official Handbook, 1914* (1914), p. 4.
[57] Derived from Archer, 'The Manchester Automobile Club Reliability Run', *passim*.
[58] *The Motor*, 11 April 1905, p. 269; 23 May 1905, p. 418; 26 September 1905, pp. 190, 212; 10 October 1905, pp. 227–8; 31 October 1905, p. 317.
[59] *The Motor*, 26 December 1905, p. 603.

Customer experience

Consumption and retailing are now important themes for the historian. The agency of the consumer in desiring goods now challenges the view that the Industrial Revolution, and production, drove growth.[60] Trentmann and Hahn have discussed the place of the consumer in the late-Victorian period.[61] Franz has shown that the consumer was also a key agent in honing motor traction to his or her needs and preferences.[62]

The letters pages in motoring magazines provided a forum for intending motorists to fulminate against an inadequate customer experience. Some manufacturers claimed to welcome prospects to their factories. Napier in 1903, for example, announced that purchasers were 'cordially invited to visit the works, and see their cars in every stage of construction'. Its brochure claimed there was 'nothing to conceal, but on the contrary, inspection is welcomed.'[63] Feedback suggests this was seldom the case elsewhere. An article in *The Motor* in 1904 referred to how English motor-cycle firms dealt with enquiries and orders. A New Zealand representative of a 'firm of repute' had written to twenty-three such firms, indicating he was prepared to pay cash in London for goods ordered. He received catalogues from ten but nothing from nine after six months. One wanted 'banker's references' even for a catalogue.[64] R. H. Wyeth of Leamington wrote to *The Autocar* in 1900: having determined to buy a car, he stopped in London to pay a visit to some firms advertising in *The Autocar*'s columns. He had time to visit four (but declined to name any). At the first, the manager was out, and a 'boy' was in charge, who admitted he 'didn't know much about motor cars'. At the second, another 'boy' was in charge, and the whole operation so 'laissez faire' that he left. At the third, the manager was out, but an 'intelligent workman' was present, although he was unable to answer pricing queries. At the fourth he listened to a tirade against another agent and how good his own cars were; the entire visit was most amusing. Only the third responded with a letter, and that was to say they normally expected forty-eight hours' notice for a visit. Meanwhile, he had been extended much courtesy at a manufacturer's

---

[60] Neil McKendrick, 'The consumer revolution of eighteenth-century England', in Neil McKendrick, John Brewer and J. H. Plumb (eds), *The Birth of a Consumer Society* (London: Europa, 1982), pp. 9–33.

[61] Frank Trentmann, *Empire of Things: How We Became a World of Consumers, from the Fifteenth Century to the Twenty-First* (London: Penguin Classic, 2017), chap. 1; Hahn, 'Consumer culture and advertising'; and for a discussion of recent themes in consumption, J. Stobart and I. Van Damme, 'Introduction', in J. Stobart and I. Van Damme (eds), *Modernity and the Second-hand Trade: European Consumption Cultures and Practices, 1700–1900* (London: Palgrave Macmillan, 2010).

[62] Kathleen Franz, *Tinkering: Consumers Reinvent the Early Automobile* (Philadelphia, PA: University of Pennsylvania Press, 2005). Also potentially useful, but as with most literature, covering a later period, are Kurt Möser, 'World War I and the creation of desire for automobiles in Germany', in Susan Strasser, Charles McGovern and Matthias Judt (eds), *Getting and Spending: European and American Consumer Societies in the Twentieth Century* (Cambridge: Cambridge University Press, 1998), pp. 195–222; and David Thoms, 'Motor car ownership in twentieth-century Britain: A matter of convenience or a marque of status?', in David Thoms, Len Holden and Tim Claydon (eds), *The Motor Car and Popular Culture in the Twentieth Century* (Aldershot: Ashgate, 1998), pp. 41–9; Ruth Oldenziel and M. Hård, *Consumers, Tinkerers, Rebels: The People Who Shaped Europe* (London: Palgrave Macmillan, 2013).

[63] Napier brochure (1903), with thanks to the late Malcolm Jeal.

[64] *The Motor*, 6 April 1904, p. 245.

in Coventry, who answered all his questions, did not deprecate the rivals, and thus secured his order.⁶⁵ 'M. I. C. E.' also reported visiting two 'business premises' in London where scant courtesy was shown.⁶⁶ 'H. W.' wrote to *The Autocar* in 1903 to report that nearly three quarters of 'the failures and troubles' were down to English-made cars: 'I have had practical experience both with English and foreign cars, and I must say I have found I get infinitely better treatment, and an altogether more reliable and in every way superior car, from the best foreign makers'. He continued, 'I have mentioned faults to the English makers and they have invariably "known best" and refused practical suggestions, and so long as this sort of wisdom prevails, the English car is not likely to improve much'.⁶⁷ 'F. F.' did not like cars marketed as 'English' when they were only assembled in England, because that meant nobody would accept responsibility for them. He had had that problem with such a car when the back axle broke after six months. 'I believe nothing in the motor trade I do not see', he declared. 'P. H. H.' imposed some simple tests. Avoid English cars, he said. The make should be French, and it should then have a good resale value in two or three years' time.⁶⁸ 'E. O. B. V.' wrote in 1902 to the principal firms for their catalogues, and then followed this up by writing to the doctor identified in a testimonial. It turned out that the doctor had given up on that brand, tried another, given up on that, and now recommended another brand altogether.⁶⁹

A broad range of experience and luck was recorded, for example from H. R. Beckett who bought an 'Ideal' Benz in 1899 and had covered over 1,000 miles with repair costs of only 4s 6d. 'I acknowledge myself to be of the "handy-man" type possessed of a fair collection of the more common mechanical tools, of which I am more or less the master', he said. Sam Wright, on the other hand, bought a Pennington and Baines in October 1898 which proved dreadful, and described the series of letters between him and the company until February 1900 when he issued a writ for recovery of his money.⁷⁰ 'An owner of three cars' wrote in to point out the number of cars offered for sale second-hand. Being offered second-hand 'means that people have been disgusted, and that the cars in many cases have been [mis]represented'. He suggested firms should make their demonstration cars available for a fee, with driver, for seven to ten days. As it was, firms often gave their trial on a paved road, secured a deposit and balance before delivery, to be inevitably followed by disappointment of the customer. 'Commercial' agreed. A firm had not allowed him to hire a car and driver with a 'trivial' excuse and this 'proves it would not do the work, nor go up the hills they guaranteed it would'.⁷¹ A letter from G. Spindler related his ordering of a car from a 'well-known' firm in June 1899 and despite letters saying it was nearly ready, he still had not taken delivery of it a year on. Delivery had been guaranteed within four months, at which point he visited

---

⁶⁵ *The Autocar*, 20 January 1900, pp. 60–1.
⁶⁶ *The Autocar*, 3 March 1900, p. 214.
⁶⁷ *The Autocar*, 31 October 1903, p. 550.
⁶⁸ *The Autocar*, 18 March 1905, pp. 390–3; 25 March 1905, p. 442.
⁶⁹ *The Autocar*, 18 March 1905, pp. 390–3.
⁷⁰ *The Autocar*, 10 February 1900, pp. 144–5.
⁷¹ *The Autocar*, 24 March 1900, p. 289, 31 March 1900, p. 311.

the factory and found manufacture had not even been started.[72] 'One of the public' described how salesmen denigrated other brands, and how his offer to pay two guineas for a fifty-mile trial in any of 'three or four' firms met with a 'contemptuous refusal'; a trial after purchase; and no replies from the rest.[73]

Even where happier experiences were recorded, it is possible some of the writers were 'plants'. 'O31' said he spent two days at the 1904 Crystal Palace Motor Show and was able to whittle his selection down to eight. He wanted a British car if possible, and one made locally [Birmingham] too, so he could get spares. Trials for him were useful for the all-round improvement of motor cars by testing silence, dustlessness, low fuel consumption and brakes. 'L. B. C.' was influenced partly by the reliability trials and partly by friends. He wanted an English car but did not think the trials were 'run on the best lines'. 'G. E. W.' also wanted a British car from a firm of good commercial standing. He was influenced by general appearance, by letters in *The Autocar*, but did not like reliability trials. 'B. W. V.' said, 'I don't believe these trials are of any use at all. Only a "juggins" or a beginner in motoring could be influenced by them'. Winding up the debate, *The Autocar* noted that an otherwise reliable car would be placed low down on the list of reliable cars because of something as unfortunate as a fouling spark plug; and on the other hand there are suspicions that special machines were entered which then did particularly well.[74] But most telling of all was the testimony of 'W. P. D. B.', a doctor with two cars. He used one in the mornings and the other in the afternoons, because he did not like dirty cars. He took 'disinterested expert advice', but thought 'the Automobile Club (of which I am a member) does not hold disinterested trials. It is run by a little trade ring'.[75]

The Motor's Trial, 1903

The letter from 'W. P. D. B.' was one of several in the motoring press expressing a clear sense amongst consumers that the Automobile Club was part of a motor-traction junta. Certainly, the Automobile Club, in cooperation with the SMMT, had achieved a monopoly over motor-show management, all motor sport, as well as advising on government policy.[76] To this end, national trials were organized mainly under the auspices of the Automobile Club, or were held as local events by clubs affiliated to the national club. There were other forums, such as the *Daily Mail* with its motor-car-promoting owner Alfred Harmsworth; this organized trials, such as the 100-miler in 1899. On this occasion, a certificate was awarded to the only British-engined car – Edge's Number 8, once a Panhard and now with a Napier engine – and the results were useful to potential customers for revealing overall speed and climbing ability.[77]

---

[72] *The Autocar*, 23 June 1900, p. 603.
[73] *The Autocar*, 23 June 1900, p. 604.
[74] *The Autocar*, 25 March 1905, p. 492.
[75] *The Autocar*, 18 March 1905, pp. 383–93.
[76] Brendon, *The Motoring Century*, p. 88. For SMMT, see Nixon, *The Story of the S.M.M.T.*, esp. chap. 3.
[77] Report on the *Daily Mail* 100-Mile Trial: *Daily Mail*, 12 October 1899.

Figure 4.3 Charles Jarrott and S. F. Edge pose on the 6hp Panhard-Levassor (known as Number 8) which had won the 1896 Paris-Marseilles race. Harry Lawson bought and imported the car before selling it on to Edge, who instructed Montague Napier to make several improvements, including here, a steering wheel replacing the tiller. Image: Peter Card.

It remained unusual, though, for publications to stage trials, but this is what *The Motor* attempted to do in 1903. The magazine had set itself up as the champion of the 'modest motorist', and that made it distinct from the opinion appearing in the more upmarket magazines which appealed to a more traditional, clubbable reader.[78] *The Motor* had always claimed to take a fearless and provocative stance, initially for the benefit of the motor-cyclist and then for the light-car owner. In its 'Opinion' piece of its very first number, it thundered, 'There is one point we must particularly emphasize – that "Motor Cycling" [a year later, it had become *The Motor*] will be absolutely unfettered in its opinion and in its policy. We shall not hesitate to speak out when speaking out is necessary in the interests of sport, pastime or trade, and the cause of the motor cyclist will always be our first consideration'.

Taking this high-minded stance, on the side of the overlooked 'modest' motorist who was probably unable (or unwilling) to secure membership of the clubs, *Motorcycling and Motoring* was critical of the Automobile Club, its *Journal*, and how the club had failed to recognize the increase in popularity of the motor-cycle in the previous few years.[79] It pointed out that the Automobile Club was powerful but 'it is the club for the more wealthy sections of motorists, and this one fact has served to

---

[78] Described in Horner, 'The emergence of automobility in the United Kingdom', pp. 56–75.
[79] *Motorcycling and Motoring*, 22 October 1902, p. 174.

make it somewhat exclusive'. *The Motor* was more generous in its opinion of other national clubs associated with the Automobile Club, such as the Auto-Cycle Club, catering for the motor-cyclist, and the Motor Union, for the 'motorist of moderate means' and established two or three years earlier. It thought the Motor Union had a great future, and 'commends itself to motorists of all grades and of every status'. *The Motor* approved of the Motor Union's more moderate subscription (a guinea a year), and its popular constitution, being unimpressed with the Automobile Club's high subscriptions – initially four guineas per year[80] – limited membership, expensive club premises in an inappropriate location, and a programme of events appealing to the few. Bemoaning the Automobile Club's minor interest in motor-cycles and light cars, *The Motor* also provocatively suggested that if the Automobile Club was too busy to attend to the needs of motorcyclists, it should relinquish 'its claims to control of motorcycles and other light vehicles' and another outfit be formed to manage this. The Automobile Club, *The Motor* said, was now keen that 'every owner of a mechanically propelled motor vehicle who is not disposed, for one reason or another, to join the Automobile Club, or one of the local affiliated clubs should be induced to join the Motor Union'.[81] Indeed, the Automobile Club and the Motor Union had an uneasy relationship until eventually the latter was absorbed into the rival Automobile Association.[82]

So when *The Motor* attempted to set up its own light car trials in 1903, feathers were ruffled. Noting just how irrelevant the Automobile Club was for representing 'modest motorists', the magazine announced its 'Run round London for light motor cars', to be held on 14 November. The run was due to start at Bromley, Kent, from where it would describe a circular route around London of twenty-mile radius, clockwise, covering about 130 miles and finishing in Barnet, Herts. This was not the first time a magazine had mooted the idea of trials for light cars: C. W. Brown in *Bicycling News and Motor Car Chronicle* had raised the need for 'motor competitions for "low-powered cars"' as early as 1900,[83] but *The Motor* was the first to organize one. It justified its trial by declaring that the light car should be the backbone of the industry, but was not getting the attention it deserved. 'The attempt [by the Automobile Club]', it said, 'to mix small cars with large cars has not been attended with any degree of success'; so, only vehicles 'for the man of moderate means' would be permitted.[84] *The Motor* had already enjoyed ribbing the Automobile Club for its poorly organized trials, the latest of which had been its reliability trial for motor cars in September 1903.[85] This particular trial had covered over 1,000 miles, was based at Crystal Palace and involved long return runs each day. Edge had been an honorary marshall, and Charles Jarrott and Dorothy Levitt were there as well.[86] *The Motor* deemed this trial to be suited to large cars only and of little value for testing light cars. It reasoned that 'it is practically impossible

---

[80] Nixon, *Romance Amongst Cars*, p. 59.
[81] *The Motor*, 26 December 1905, p. 587; 11 March 1903, p. 353, article on the Motor Union; *Motorcycling and Motoring*, 10 December 1902, p. 326, opinion piece; *The Motor*, 24 June 1903, p. 460.
[82] Brendon, *The Motoring Century*, p. 88.
[83] *Bicycling News and Motor Car Chronicle*, 13 June 1900, p. 14.
[84] *The Motor*, 21 October 1903, p. 251A–B; 'Light motor car run round London'. The trial was touted heavily over the following weeks: 4 November 1903, p. 295.
[85] *The Motor*, 23 September 1903, pp. 153–9, 30 September 1903, p. 175, 7 October 1903, p. 207.
[86] See *The Motor*, 23 September 1903, pp. 153–9, 30 September 1903, pp. 175, 177–84.

to frame a set of conditions or scheme of comparison' that could meaningfully compare a large car with a light car. *The Motor* offered the barb that the club was now uncharacteristically very well organized, in a dig about just how poor some of the Automobile Club's other efforts had been to date. In *The Motor*'s opinion page of 4 November, '[our light car run] will prove to be not only the event which celebrates the independence of the light car, but will attract the attention of thousands to whom it never before occurred that motoring was otherwise than something quite beyond their means'.[87] The magazine claimed to have received solid support from 'firms in the trade' and identified eleven different makes that had offered entries, and 'expected' entries from another thirty-one.

The other motoring-magazines feigned horror. The very appearance of *The Motor* only eighteen months earlier had irritated the existing magazine editors, manufacturers and agents, for whom advertising budgets would now need to be spread more thinly.[88] Referring to *The Motor*'s 'unofficial trials', *The Autocar* decided that the Automobile Club should remain the only organization to organize trials because otherwise 'there is no knowing where the thing will end'. 130 miles was no test, it cautioned, and anyway, there were the quarterly 100-mile Automobile Club trials, for which there was a non-stop certificate. The magazine urged that the Automobile Club should disqualify all participating makers in *The Motor*'s trial from its own reliability trials, or the 'so-called reliability runs will become a nuisance to the manufacturer and the laughing stock of the public'.[89] So horrified was *The Autocar* that it suggested the trial was 'sufficiently mischievous for us to depart from our regular practice' of avoiding criticizing a 'contemporary'.[90]

Indeed, suddenly, the rug was pulled from under the feet of *The Motor*. Manufacturers wrote in to deny offering *The Motor* support. *The Motor* published an acrimonious editorial condemning the Automobile Club for its skulduggery.[91] It is unclear where 'blame' really lay for what happened next, but *The Motor* abruptly cancelled the trial, and pointed the finger firmly at the Automobile Club. The General Committee of the Automobile Club had agreed on a resolution not to support the trials, overriding an agreement of the Executive Committee. The SMMT also resolved not to support it.[92] *The Motor* accused the Automobile Club of 'inconsistency and absolute lack of sincerity', and published extracts of letters exchanged between Dangerfield and the Automobile Club Technical Secretary Basil Joy (1870–1940) to prove that news of the intended trial had been received favourably by the industry.[93] It is likely Dangerfield pulled the plug on the trial because he feared the loss of advertising revenue from those manufacturers who had pulled out, but organizing an event so hastily may also have meant its cancellation was silently to be welcomed.

---

[87] *The Motor*, 4 November 1903, p. 295.
[88] See the letter of William Letts of the SMMT to Dangerfield explaining this. Iliffe, publisher of *The Autocar*, used the shift in *The Motor*'s direction to launch *The Motor Cycle* in 1903: Armstrong, *Bouverie Street to Bowling Green Lane*, pp. 71–2.
[89] *The Autocar*, 24 October 1903, pp. 506–7.
[90] *The Autocar*, 31 October 1903, p. 533.
[91] *The Motor*, 11 November 1903, p. 326; 18 November 1903, p. 358.
[92] *The Autocar*, 7 November 1903, p. 589.
[93] *The Motor*, 11 November 1903, p. 326; 18 November 1903, p. 358.

The Motor continued to fume at the Automobile Club's shenanigans, which in turn recognized it now needed to stage a light-car trial at its next quarterly trials in 1904. The fee charged by the Automobile Club for each entrant was high (ten guineas per car, of which six was returnable if the car turned up), at which many of the smaller manufacturers would have blanched.[94] But this was an opportunity for a manufacturer or agent to earn an Automobile Club certificate for a one-hundred-mile non-stop run, and therefore earn advertising copy. *The Motor* objected, 'The trade set[s] very little value on the quarterly trials' – 'from the public point of view, [. . .] practically nil'. The trials are 'carried out in obscurity' and rarely does a record appear in the public press. *The Motor* concluded, 'The quarterly trials are dead to the public'.[95] But the Automobile Club went ahead, stealing the magazine's thunder, dedicating its annual reliability trial entirely to light cars.[96] This became the Hereford event that Levitt participated in, which, while having non-stop runs daily, was less rigorous than its usual trials. Again, in recognition of the more modest specification of the entrants, speeding was discouraged, and anyone averaging over 18mph was disqualified. Its object was 'to demonstrate to the purchasing public the capabilities of the light car to-day, and to guide them in the choice of a reliable vehicle'.[97] Its 'real-world' tests included an unannounced brake test and restart on a 1-in-7.8 hill. (Most passed.) The Siddeley, with twelve non-stops, then advertised its triumphs in *The Motor* and elsewhere,[98] meaning that if *The Motor* could not get to hold its own trial, it could at least derive some advertising income.

Conclusions

Public trials of motor-cycles and motor cars rapidly became important opportunities for agents and intending customers to come together. Agents could now parade and demonstrate their products, and any success in a trial meant positive copy for their advertising, while customers could examine the goods at close quarters, see, hear and smell them, and speak to the drivers and officials. For the agent, though, providing vehicles for trials was always a risk because of the possibility of breakdown or accident, while road conditions meant vehicles arrived unavoidably filthy. It was also very expensive for the early manufacturers and agents to supply vehicles, because a sum of money, perhaps as much as ten guineas, was needed upfront for each entry, although this was necessary for the organizers to minimize no-shows, and some of the fee was returnable if the vehicle actually turned up for the trial.

Trials soon became increasingly stringent. First of all they tested for reliability, hill-climbing or braking, but within a few years became showcases for, say, driving entirely in top gear from one end of the country to the other (gear-changing was the bane

---

[94] *The Motor*, 30 December 1903, p. 545.
[95] *The Motor*, 30 December 1903, p. 545.
[96] *The Motor*, 7 June 1904, p. 487. Discussed earlier in the chapter; see Brooks, 'The Small Car Trials of 1904', pp. 18–39.
[97] *The Motor*, 21 June 1904, p. 578; 6 September 1904, pp. 119–27, with editorial on p. 128.
[98] *The Motor*, 21 June 1904, p. 578; 13 September 1904, ads.

of the driver's life, and so a car that could go uphill and down dale entirely in one gear was seen as attractive). The trials variously informed, bemused and infuriated the consumer, particularly when there was any discrepancy: it was not unusual for a vehicle to be awarded a prize but, on scrutiny, that vehicle may have endured significant breakdowns or failures – letters show that readers did not fail to spot this. The results could also make it difficult to differentiate between a breakdown due to, say, a puncture, or the failure of a major component, such as a broken axle. But trials did serve the purpose of bringing together vehicles for public scrutiny and information exchange. They in turn provided an opportunity for product development for the manufacturer; they brought glamour, photographic opportunities and copy for the magazines; and created the advertising sales 'puff' for the agent, particularly if the product did well.

The local clubs, all affiliated to the Automobile Club, also played their part in presenting vehicles for public scrutiny in managed events such as reliability runs and hill-climbs. Local motor clubs – and here the Manchester Automobile Club and Manchester Motor Club have been used as examples – were formed and run in the vein of (horse-)driving and cycling clubs, with a president, a captain, a rule-book and a full social calendar. They too sought to ingratiate themselves with the influential, such as chief constables, MPs and lord mayors, making sure all were invited to their presentations, dinners and events. The clubs were aware of the need to showcase motor traction as a sensible and safe activity, and built in minimum and maximum times for stages of their reliability runs which ensured drivers could not race. The kicking up of dust, and the potential to cause a nuisance, though, was always present. However, any local event presented opportunities for the public to mill around, browse and enquire, thus removing much of the mystique for the uninitiated.

The Motor's hapless attempt in 1903 to put on its own trials was an opportunity for the national club, the Automobile Club and the collective of agents and manufacturers, the SMMT, to assert control and tighten their grip on who could officially test and exhibit motor cars in the public gaze. *The Motor* was obliged to abandon its intended event – a circular tour of London for light cars only – probably under pressure from advertisers and agents. However, the cancellation of the event meant the Automobile Club was ultimately compelled to stage a trial shortly afterwards, also specifically for light cars, and it did so in the following autumn (1904). This trial was conducted with a less exacting schedule than would have been used for bigger and more expensive motor cars in line with what *The Motor* had wanted for its own trial. Also, the clash, while in the short term a humiliation for the magazine, showed that *The Motor* was more in touch with – or, perhaps, prepared to get to grips with – evident changing consumer patterns. As will be seen, *The Motor* had positioned itself to speak directly to the middle-class intending motorist, and therefore in the main 'below' that of *The Autocar*. It had found its own voice, then, resisting being another mouthpiece for the Automobile Club, and offering a clear break from the world of the clubs and cliques.

The place of *The Motor* magazine is significant, as the next chapter will show. Its creation of the 'modest motorist' showed it was sensitive to the consumers' needs, creating as well as able to respond to new patterns of demand, evolving from cycling, to motor-cycling, then light cars, and then spinning off a cyclecar magazine. Its new appeal was apparent by its endorsement of the practice of buying second-hand, by its discussion of cheaper cars, of hinting at lower running costs and by a tacit endorsement of buying 'on tick'. *The Motor* also recognized the emergence in the early years of the twentieth century of a new kind of consumer, one who had possibly never cycled, who had scant interest in the mechanicals, who was more conscious of changing styles and who wanted convenience and reliability while having an adventure. The same consumer was probably not clubbable but then did not seek club membership. It is the emergence of this consumer and his or her influence on the shaping of the emerging product that forms the subject of the next chapter.

5

The 'old brigade' and the new 'steady and careful artisan'

In late summer 1897, Charles Jarrott was asked by pioneer motorist Charles McRobie Turrell (1875–1923) to collect the ex-racing Panhard-Levassor motor car (called Number 5) from Margate, and deliver it to the offices of Harry Lawson in London. Jarrott asked Frank Wellington (1868–1917) to accompany him, as Wellington was an 'authority' on motor cars, and had been motoring since before the 'Emancipation Run'.[1] Once on the road, it took two hours of 'winding' the starter handle to restart the car. Jarrott described how, thereafter, the journey became increasingly miserable as darkness fell, and the 'crowning disaster' was when their candles – the only means of seeing ahead – gave out.[2] They travelled all night, arriving in London the following morning.

There would not have been anybody that Jarrott could have consulted to be shown the 'right' way to start the vehicle as even the manufacturers were in the dark as to the method. David Salomons (1851–1925) recalled the same experience. In the 'early days' (he wrote in 1906), there was little or no information to be obtained; all the pioneers were self-taught. He remembered that the manufacturers would demonstrate what they thought was the best way to start the motor vehicle. He went to Paris to collect his Peugeot and two half-hour runs with an engineer was all the instruction given, after which he was let loose in Paris all alone. On this, and other early cars, 'I made my apprenticeship, and learnt a great deal in a very practical manner, buying experience expensively with time, money, vexation and many flesh wounds.'[3]

Henry Sturmey also reflected on the earliest experiences. He wrote about the period 1902-3, 'Those were the days . . . when people were still a little bit afraid of the idea of taking a car out by themselves, for fear it should "break down", and that they would not be able to get it back again, and when such timid ones were gravely told that to get decent satisfaction out of a car it was necessary that a good chauffeur should be employed, at a salary of £5 per week'. However, Sturmey acknowledged a fundamental

---

[1] https://gracesguide.co.uk/Frank_Frederick_Wellington, accessed 14 May 2020.
[2] Jarrott, *Ten Years of Motors and Motor Racing*, pp. 13–18. Wellington became the agent for Phebus-Aster and Brooke motor cars.
[3] David Salomons in [Lord] Montagu, *The First Ten Years of Automobilism*, p. 4.

shift in attitudes and expectations towards motor traction. He was writing in 1914 when he said:

> The field of motoring is now opened, not only to the rich and well-to-do, as in former years, but also to the bulk of the middle classes, and is not even limited to these, as the cheaper forms of motor vehicles are now within the reach of the steady and careful artisan, indeed, of all who can get together a few pounds for purchase, and are content to limit their motoring to their income.[4]

Motorcycling magazine also looked back for its readers, noting how the motor-cycle was now reliable and did not vibrate: 'nowadays [1910] a man need not be an expert mechanic to run a motor-bicycle successfully, neither need he don ultra-weird garments to go for a ride'.[5]

Clearly, something had drastically changed in the few years since Jarrott's miserable journey to allow motor traction to move from a niche, hobbyist activity pursued by a wealthy enthusiast, to one of appeal to a wider, non-technical audience. This was a process driven by the entrepreneur, putting on and competing in trials, setting up magazines, creating brands, putting an infrastructure in place, and constantly reminding the consumer of the rapid progress being made, all this creating the desire to go motoring. It took the tinkering of entrepreneurs to develop a product more convenient to use, and therefore easier to sell. Edge, for example, had found the tiller steering on his Panhard Number 8 'quite hopeless', so switched it for a steering wheel.[6] But on the other hand, there was also clear input from the consumers, voicing indignation in the motoring magazines about inadequate products, telling the entrepreneurs what they wanted in terms of vehicle specification and size, what they expected for a given price and what they wanted to learn from trials. This chapter, then, will consider the rise of motoring as an increasingly normalized activity from the points of view of both entrepreneur and consumer, as the motor car and motor-cycle moved from hobbyist freak shows to viable, standardized (in terms of general form and technical configuration), attractive and (almost) sensible consumer goods, all in the space of a generation. It will suggest that the first tier of consumers (the pioneers, or Jarrott's 'old brigade'[7]) was eclipsed by a much bigger group – Sturmey's 'steady and careful artisan'.[8]

Diversity and standardization

The appearance of the 'safety' bicycle, with its diamond-shaped frame, in the 1880s, presaged a standardization of bicycle design, still evident today. It enabled non-sporting

[4] Henry Sturmey, 'The continued growth of popular motoring', *The Motor*, 31 March 1914, pp. 350–2.
[5] *Motorcycling*, 28 June 1910, p. 190.
[6] Edge, *My Motoring Reminiscences*, p. 62. There is a photograph of Edge aboard the unmodified Panhard in *Veteran and Vintage*, 4:3 (1959), p. 452.
[7] Charles Jarrott, 'By way of introduction', in Jarrott, *Ten Years of Motors and Motor Racing*, n.p.
[8] Sturmey, 'The continued growth of popular motoring'.

men and women to cycle, and undermined the exclusive appeal of the 'ordinary', which, in the 1870s and 1880s, had been accessible only to athletic, middle-class, risk-taking (and long-legged) men.[9] Indeed, such exclusivity had been part of its attraction. From the late-1880s second-hand 'ordinaries' appeared increasingly on the market. Not before the 1890s, then, could the bicycle, in its 'safety' form, be mass-marketed.

A similar 'standardization' was necessary for the motor car: while the 1901 Mercedes is given as the exemplar that moved motor traction on from being little more than a horseless carriage,[10] it was not until about 1905 before it had gravitated to a (mostly) common standard (cars having wheels the same size, engines at the front driving the rear wheels; motor-cycles having engines mounted low down between the wheels), particularly for light cars. Until then, there was no 'standard' product to draw in a wide, non-technical 'consumer' and motorized mobility remained the preserve of the clubman and the 'hands dirty' enthusiast. That same standardization had brought about a tolerable ease of starting and reliability, and therefore a likelihood of actually arriving; this also raised confidence amongst prospective customers in the product. (Of course, some of the older, pioneering generation were less inclined to move on wholesale. Club cyclist Montague Napier, for instance, continued to ride his 'ordinary' into the 1900s while simultaneously managing his state-of-the-art motor-car company; and racing driver Charles Jarrott bemoaned as early as 1906 how 'the romance of motoring exists no longer' now that motor cars had become demonstrably more reliable.)[11]

As with the 'maturity' of the bicycle, for motor traction it was a slow and haphazard process. While 'early adopters' of the bicycle, motor-cycle and motor car had been game for all manner of technical approaches, it was precisely this that kept the non-technical and non-sporting at the wayside. A cartoon in *Motorcycling and Motoring* magazine in 1902 revealed that it was this very uncertainty that was holding consumers back. Picturing the 'Agonising nightmare of Jones after a visit to the Crystal Palace Motor Show', it suggested 'The best position for an engine!', mimicking the likely sales patter from the different salesmen with their varying contraptions to sell, and the conflicting dozens of positions in which the engine might then have been mounted in a motor-cycle.[12] It is perfectly obvious now, of course, to have the engine of a motor-cycle mounted between the wheels and low down, but that was simply one possible position of many in those early days. For buyers of smaller motor cars, too, there was not then any standard in engine or seating position, nor even a recognized 'look'. As an indication of the variety on the market, motor-car buyers in around 1902 could choose from, amongst others, the Sunbeam Mabley for £140,[13] with its four wheels in diamond

---

[9] Ritchie, *Early Bicycles and the Quest for Speed*; Reid, *Roads Were Not Built for Cars*; Andrew Ritchie, *King of the Road: An Illustrated History of Cycling* (London: Wildwood House, 1975). As an example of the exclusivity of the 'ordinary', Edmund Dangerfield, subsequently editor of *Cycling* and *The Motor*, notably short in height, was therefore understandably not known to have ridden an 'ordinary', but went on to win the '100' [100-mile race] for the Bath Road Club in 1890 on a 'safety': Armstrong, *Bouverie Street to Bowling Green Lane*, p. 10.
[10] See Anthony Bird, *The Motor Car, 1765–1914* (London: B.T. Batsford, 1960), esp. pp. 84–7.
[11] Vessey, *By Precision into Power: A Bicentennial Record of D. Napier and Son* (Stroud: Tempus, 2007), p. 46; Jarrott, *Ten Years of Motors and Motor Racing*, p. 266.
[12] *Motorcycling and Motoring*, 5 March 1902, pp. 64, 70–1.
[13] *Motorcycling and Motoring*, 3 December 1902, p. 298.

Figure 5.1 S. F. Edge remained a cyclist all his life and was happy to endorse products, here an Ariel, 1909. From G. H. Smith, *Selwyn Francis Edge: The man and some of the things he has done* (privately pub., 1928), p. 12.

formation and the driver sitting at the back, having to look around his passengers; or the Carpeviam, a German import with three wheels, of a type 'in great demand by commercial and professional men, who can neither afford the price of a large motorcar, nor do they intend to carry much dead weight on their daily rounds', and seen by *Motorcycling and Motoring* as thoroughly reasonable at £99.[14] Lawson's Gyroscope of the same year was little more than a trap with the horse replaced by an engine mounted atop a steered wheel – he arranged for his daughters to be photographed piloting one, to suggest the ease of use. The journalists and entrepreneurs of the day saw great potential with this arrangement, and the magazine asked, rather hopefully: 'will the Motor, adapted to any kind of carriage become a feature of the future?'[15] Here, the writer was speculating that perhaps clamping a steered, powered wheel literally

---

[14] *Motorcycling and Motoring*, 5 November 1902, p. 204; 17 December 1902, p. 339. The Carpeviam is illustrated and described in 5 November 1902, p. 204.
[15] *Motorcycling and Motoring*, 12 February 1902, p. 17.

in place of the horse might soon be 'normal'. *The Motor*, meanwhile, celebrated this very diversity with a double-page picture of 'Men of moderate means out motoring', which included a Mabley.[16] 'V. E. B.', a GP from Devon, summed up the bewilderment, wondering in *The Motor* if he should spend his £160 budget on a steam- or petrol-powered motor car, and whether it should be new or second-hand, and with solid or pneumatic tyres.[17] Enquiries such as these led to editorial responses, such as that by 'Automan' offering advice on 'The selection of a car'. He asked, 'What is the car required for?' – business or pleasure? Do you live in a hilly district (in which case the car must have hill-climbing ability)? Do you intend to keep a 'mechanicien' to look after or drive it, or do it all yourself? (He pointed out, if it were a large car, you could only really look after it yourself if you were 'of great leisure'.) He explained how petrol was by then clearly the most popular means of propulsion, but electric and steam vehicles were still available, especially second-hand, and might be more suitable for town work. He asked, do you want your car for general and touring purposes? Should you opt for a two- or four-seater body?[18] This was all very sensible for the consumer who had no idea where to start.

As the Carpevian, Sunbeam Mabley and many others suggest, there was initially no technical standardization for smaller motor cars. It was fully understood that the motor-cycles and motor cars were a handful to master, both temperamental and underdeveloped. 'Knowing' articles, intended to amuse, appeared, working on that very basis, such as 'How to wrestle with a quad: By a beginner for beginners',[19] but which also played a key role in informing about elementary servicing and repairs. The enthusiast played a key role here. The home-made contraption continued up to and beyond the First World War, as users found they could build a product cheaper, better or just different to what was on the market. The Cheshire motor vehicle registrations bear this out, with a handful of motor-cycles and other larger vehicles registered without a brand name, or registered with the surname of the owner, and therefore in all probability assembled in the owner's back yard. 'A. J.' of Lincolnshire bears this home-built approach out. In his letter to *Motorcycling and Motoring* in 1902, he wrote, 'I thought a 3-wheeler could be built more cheaply than a 4-wheeler, so I built a 3-wheeler on the Century pattern', that is, with one rear-driven wheel. However, he found it skidded more, so that wet tram lines were a 'terror' – while in the wet, the back wheel slid into the gutter, and at more than 6mph on wet roads it was not safe. His improved version had a long wheelbase, a 4hp engine, much as per the Carpevian. It was made to carry three, but often carried four, did 16mph, although rear-tyre wear was tremendous. If he did it again, he said, he would make it a four-wheeler that could run on wet roads safely.[20]

In offering solutions on improving 'light cars', a DIY culture flourished. Letters were published through to 1914 and beyond by people intending to build their own light car.

---

[16] *The Motor*, 25 March 1903; the image is reproduced in Horner, 'The emergence of automobility in the United Kingdom', p. 58.
[17] *The Motor*, 1 July 1903, p. 497.
[18] *The Motor*, 14 November 1905, pp. 355–6.
[19] *Motorcycling and Motoring*, 7 January 1903, p. 403.
[20] *Motorcycling and Motoring*, 17 December 1902, p. 356.

Figure 5.2 Harry Lawson played a key part in promoting the early motor industry, and was behind the London to Brighton Run of 1896. Here is his 'Gyroscope' of c. 1902, with his daughters demonstrating its apparent ease of use. With the horse directly replaced by a steered, motorized wheel, commentators wondered at the time if this approach was the direction motor traction would take. From H. O. Duncan, *The world on wheels, vol. 2* (Paris: privately pub., 1926), p. 815.

In most efforts, there was probably no intention by the owner-builder to move to wider production, but sometimes there clearly was. 'Atlas', for example, in 1903, wrote to *The Motor*. Under the heading 'The type of car that is wanted', he pondered the recently featured Morette, with its front single, driven, wheel, and bath-chair arrangement for the passenger. He concluded this was inadequate because of vibration and the inability to stop in traffic. He had designed a 4hp three-wheeler for £65, as well as a 6hp four-wheeler for £90, but now needed capital to develop them for production, and sought help for this.[21] This was one means by which bright back-shed inventors could draw themselves to the attention of potential backers.

---

[21] 'The possibilities of light motor vehicle design', *The Motor*, 11 March 1903, p. 94. The Morette is illustrated in *The Motor*, 25 March 1903, p. 154. Related correspondence on 22 April 1903, p. 247; 5 August 1903, p. 637, in which the 'improved' Morette 2½hp is 70 guineas, and the 4hp 80 guineas.

To whet the consumer's appetite, the magazines saw benefits in pointing out year-on-year advances. Just 'a year or two ago', *Motorcycling and Motoring* magazine reminded readers in 1902, the motor car 'was dreadfully behaved – it frightened old ladies, policeman hesitated to hold one up in case it exploded'. But now, its behaviour had become 'exemplary'.[22] Advertisers were commending the mastery of the essentials, such as ease of use, starting up, driving, going up hills and stopping. The motoring press and the *Daily Mail* saw advantages in promoting a particular sub-set of motor traction, probably in liaison with entrepreneurs. For example, as early as 1899 the *Daily Mail* was asking rhetorically, 'shall I buy a motor cycle?' The writer claimed to have ridden a motor tricycle every day for three months, covering up to 120 miles in a day, and 'nothing has gone wrong with it'. To emphasize its ease of use, he continued, 'As a matter of fact, there are thousands of lady motor cyclists abroad, and one or two in England. A person who cannot manage a motor cycle is not gifted with ordinary intelligence.'[23] In the same newspaper Edge pointed out that when 'a motor-bicycle' will take its rider a hundred miles at the cost of a shilling, 'there is not much to fear as to the future of automobile locomotion'.[24] The *Daily Mail* had already commended the utility of the motor-cycle, and observed that on the London-to-Portsmouth road that previous spring Sunday, there were almost as many out as cycles. They were seen attached to trailers, towing cycles and tandems, even towing stricken motor-cycles. *Motorcycling and Motoring* had published a drawing of a motor-cycle pulling a bath-chair trailer and two prams, with wife, three children and luggage in tow.[25] 'Names' from the trade, such as Jarrott and Edge, were commissioned to write articles for the motor-cycling press to encourage take-up. 'Why I prefer motor cycling to any other form of motoring', was one of Jarrott's articles, while Edge's was entitled, 'Why is the motor bicycle bound to become popular?' 'The motor bicycle has most of [the] advantages [of the bicycle]', he explained, 'and one greater than all – it gives to the weak or the strong advantages that the most perfectly trained cycling athlete even has not got, that of being able to run continuously at speeds that only the most perfect specimen of mechanical traction can accomplish'. It was cheaper per mile than the train, and 'infinitely quicker', as it 'takes him from his door to any destination fancy wills him'.[26] 'Cylinder' in *The Motor* played up the distinct advantages of the forecar over the motor car. Cost was one, at about £60 – a motor car would cost twice as much and be more troublesome. He continued, the forecar could be used as a two-wheeler and thus as a bicycle. One man can get a forecar up the steps through the front door and it can be kept in the hall. You do not need a 'motor house or mechanic'. If it broke down, it would be easy to fix and anyway, it could be pedalled to the nearest station. It could be stowed in the luggage compartment of a train without requiring special rates, and if you had a collision, the worst you would do was buckle your wheel.[27]

---

[22] *Motorcycling and Motoring*, 25 June 1902, p. 326.
[23] 'Shall I motor cycle?', *Daily Mail*, 12 October 1899.
[24] 'Motor notes', *Daily Mail*, 10 April 1903.
[25] 'The holiday season: True economy suggests this sort of thing to the motor cyclist with frugal mind', *Motorcycling and Motoring*, 23 July 1902, p. 389.
[26] *Motorcycling and Motoring*, 26 February 1902, p. 46 (Jarrott); 9 April 1902, p. 145 (Edge).
[27] *The Motor*, 2 September 1903, pp. 74–5.

Figure 5.3 Montague Grahame-White bought this Leon Bollée second-hand for fifty pounds in 1898. Charles Jarrott is in the driving seat, and Charles McRobie Turrell is standing behind, in a peak cap. From Montague Grahame-White, *At the wheel ashore and afloat* (London: G.T. Foulis, n.d. [c. 1935]), opp. p. 31.

The forecar turned out to be a fad. In fact, it was not even clear to the consumer (or anyone, for that matter) whether the motor-cycle might even be just a passing phase. Alluding to the maturity of the bicycle, with the diamond-frame now 'normal', 'Cyclomot' reminded readers that 'much the same sort of thoughts must have worried the lover of the "good old ordinary" when the first safety [bicycle] came home', finishing rather hopefully, 'much the same thoughts will give us pause when our first flying machine is purchased'.[28] In their moment, some really did believe this, as seen in Chapter 7.

Motorcycling and Motoring claimed to spot a pattern for a 'standard' technical specification for the light motor car by 1902, to include a tubular frame, a water-cooled engine of five or six horsepower, with three speeds and a reverse, 'and if possible designed on the most up to date lines, so that silence and sweetness of running are secured. Wooden wheels with large pneumatic tyres should be fitted, and the power should be conveyed through an universal shaft and bevel gearing.' The 'less ambitious motorist', whose 'purse is small' could probably make do with 3½–4 horsepower, two speeds and no reverse.[29] The magazine turned out to be right, and a year later (as *The Motor*) it was able to report that a clear standardization of the light car was in place, now including a tubular framework, a pedal-actuated clutch, a gearbox with three speeds and reverse, and a differential on a live rear axle.[30] The essential outline of the 'modern' motor car, and which survived for much of the twentieth century, was now evident,

---

[28] 'Cyclomot', 'Transition, or only a phase?', *Motorcycling and Motoring*, 12 February 1902, p. 17.
[29] 'Light cars: Some of the requirements', *Motorcycling and Motoring*, 31 December 1902, p. 390.
[30] 'The progress of the light car', *The Motor*, 2 December 1903, pp. 427–8.

with a front bonnet and enclosed engine; the French De Dion voiturette epitomized the style, and at about £200 set the standard for that class. Here, fortuitously, Edge, Jarrott and Duncan had formed the De Dion Bouton British and Colonial Syndicate Ltd in 1899 to sell all De Dion's output in the United Kingdom; staff associated with the concern included the old cycling companions Stocks and Hubert Egerton (1875–1950).[31] Dorothy Levitt also drove one for her book *The woman and the car* (1909), finding it 'ideal' for the woman driver.[32] Meanwhile, Muriel Cusins wrote in to *The Autocar* in 1900 to say how excellent she thought her De Dion Voiturette was, with her bicycle now 'too slow, and too "ordinary" after "motoring"'.[33] (Edge knew the Cusins [Nixon] family through the Anerley Bicycle Club, and Alfred Nixon was Muriel's father. Edge presumably supplied her with the De Dion, so it might be supposed the letter was not entirely spontaneous.) Even the writer Major C. G. Matson's car was a De Dion.[34]

Trading up

This shift towards a standardized motor car requiring less maintenance was opening the way for the non-technical consumer, as reported by *Motorcycling and Motoring*:

> of course the very reason for the existence of the light car was the general desire for more ease and comfort. Motor-bicycles are all very well for your adventurous youth scorning the sybarite joys, but when one begins to grow elderly and sedate there are times when a buzzing fly-wheel between your legs rather gets upon your nerves than otherwise. Likewise, when the aforesaid young man's fancy lightly turns to thoughts of love, the quad or the bicycle with a smart trailer or fore-carriage hardly satisfies him, whilst the trailer must necessarily offend all his ideas of ardent gallantry. He yearns for a more sociable conveyance, where the lady of his heart may sit at his side. So the light car has come to stay.

These advances in the comfort of light cars, *Motorcycling and Motoring* concluded, 'may be little things, but they mean much to the public whose education in motoring is only just beginning: and if we want to get hold of the great middle-class we must begin by consulting their comfort'.[35] The 'adventure machine' described by Mom,[36] was becoming, then, rather more reliable and comfortable.

---

[31] See Duncan, *The World On Wheels*, ii, p. 812; Hubert W. Egerton (ed., Malcolm Jeal), 'Early motoring experiences', *Aspects of Motoring History* 4 (2008), pp. 4–17; *Vintage and Veteran*, 4:3 (1959), p. 450.
[32] Levitt, *The Woman and Her Car*, p. 17, and photographed throughout.
[33] *The Autocar*, 10 February 1900, and see p. 265 for a photograph of her on the same. Miss Cusins went on to enter her De Dion in the 100-mile Voiturette Trial of April 1900, but was subsequently disqualified because her passenger (Edge) was deemed partisan: *The Autocar*, 21 April 1900, pp. 371–2.
[34] For De Dion, see Michael Edwards, *The Tricycle Book, 1895–1902, Part 1* (Brighton: Surrenden Press, 2018).
[35] *Motorcycling and Motoring*, 31 December 1902, p. 389.
[36] See Mom, *Atlantic Automobilism*, chap. 2.

Some motoring entrepreneurs began to promote motor traction as a field into which (almost) anyone could aspire to enter. First-generation cyclists were encouraged into the fold. Edge had alluded to an 'increasing class of old cyclists who rode long distances without an apparent effort, but who now, either through wealth, business, more subdued age, or a combination of all three, find the ordinary [here, he means 'safety'] bicycle of commerce too much of an athletic performance'; they will be drawn to the motor-cycle.[37] In an opinion column, the 'old time cyclist' could now be a motorist. The 'old rider', it continued, can no longer keep up with youngsters, or finds pedalling 'too great a labour'. This is why 'the elderly rider is joining the ranks of the motists'.[38] The same shift can be seen for motorcyclists. G. S. Bridgman of Torquay wrote in to *Motorcycling and Motoring* to say he had been a bicyclist for thirty years, and had bought a motor-cycle at the encouragement of the magazine, but

> alas! Sixty and odd years will tell upon the nerves, and makes one desire for a somewhat less nerve-trying and more luxurious seat while manipulating the switch, air, gas and sparking lever, with the brake, horn and steering to be attended to at the same time, than there is to be found on a motor-bicycle. The motor tricycle as at present (especially for two) does not appear to be either fish, flesh, fowl, or good red herring. . . . Our manufacturers should study the middle classes and people of moderate means in their manufacture, and produce a vehicle of good English make, somewhat after the character of the 'Carpevian'.

The vehicle should be for two, with a maximum speed of 16mph, and at a moderate price.[39]

Cyclists were being persuaded that moving to motor-cycling was a complementary, natural, activity. From 1902 the use of motor-cycles as pace-makers for amateur cycle racing was allowed, simplifying the need hitherto for four- or even five-seater tandem pacing bicycles; the Polytechnic Cycling Club put on a race at Crystal Palace with the new innovation, which drew a good crowd[40] who came not just for the sport but also for the spectacle of a motor-cycle as a tireless pacer. By the end of 1902, *Motorcycling and Motoring* was giving reasons why the 'wheelman' (cyclist) will become a motorist (by which it meant a driver of anything motorized). This was because the bicycle had reached its zenith and had little scope for further improvement.[41] The cyclist was being advised that the costs were not as excessive as thought: a good bicycle might cost fifteen pounds, whereas a motor-cycle would only be £40–45.[42] Just a few months later, its editorial commented: The 'line of demarcation between the motor-cycle and the light car is remote to the point of non-existence'. In 'learning one, you practically learn the other', and soon you'll be able to buy a cheap, light two-seater motor car for the price

---

[37] S. F. Edge, 'Why is the motor bicycle bound to become popular?', *Motorcycling and Motoring*, 9 April 1902, p. 145.
[38] *Motorcycling and Motoring*, 16 April 1902, p. 145.
[39] *Motorcycling and Motoring*, 19 November 1902, p. 254.
[40] *Motoring Illustrated*, 10 May 1902, p. 250.
[41] This point has been raised by Oddy for the period from 1900 to 1920: Nicholas Oddy, 'Cycling's Dark Age? The period 1900–1920 in cycling history', *Cycle History*, 15 (2004), pp. 79–86.
[42] *Motorcycling and Motoring*, 19 November 1902, pp. 242–3.

Figure 5.4 Typical of magazine imagery of the time, George Moore's motor-cycling couple were oblivious to the foul weather. Moore had hitherto been a cycling-magazine illustrator, and was on the staff at Temple Press, the publishers of *The Motor*. Image: *Motorcycling*, 27 December 1909, cover.

of a 'high grade motorcycle'. 'The motorcyclist is a motorist . . . his interests are entirely the interests of the great fraternity of motorists'.[43]

Making the leap from motor-cycle to motor car may have been too big for some, so an increasing selection of after-market accessories, costing money but less than the transition, could bring to the motor-cyclist the shared motoring experience

---

[43] *Motorcycling and Motoring*, 28 January 1903, p. 469.

(companionship, being able to have a conversation on the move) that the motor car brought. *Motorcycling* magazine had as its cover for a winter edition a drawing of sensibly dressed motorcyclists, one of whom had attached a sidecar whose female passenger was cosily cocooned: 'Sport all the year', it trumpeted.[44] An advert for a sidecar declared, 'your motor bike is an unsociable thing for one. Attach it to a Liberty Side-Carriage and have a really sociable vehicle for two. The cost isn't great. . . . Your motor bicycle becomes practically a small car at very much less than small car price'.[45]

The Motor

One particular magazine stands out for its attempt to differentiate the potential 'ordinary' motorist from the clubby elite. *The Motor* magazine had started out in 1902 as *Motorcycling and Motoring* with appeal for the middle-class owner of motorcycles, that is, with a wider social remit than some other magazines. Its roots were obvious: as a spin-off from its sister magazine *Cycling*, the new magazine justified itself by saying that motor-cycles (or rather, bicycles with motors) had been displayed at cycle exhibitions as far back as the 1880s (notably the Parkyns of 1881, discussed in Chapter 2, but also the Butler motor-cycle first seen in 1884), and that *Cycling* magazine would continue to be advertised in *Motorcycling and Motoring*, suggesting a readership in both camps.[46] But, then asking, 'A new paper – is it wanted?', it concluded:

> In the first place, great interest is being taken in the new [motor-cycling] movement by the public. This is proved by the fact that the readers of 'Cycling' placed the motor bicycle at the head of a list of desirable innovations and gave it a good majority. [. . .] In the motor bicycle we have the cheapest, handiest, lightest and simplest power-propelled vehicle that has yet been introduced . . . to many thousands of riders of cycles the luxurious motor car is a forbidden pleasure on account of its prime cost, and the expense of maintenance. But in the motor bicycle the cyclist has a vehicle that particularly appeals to his fancy, and his pocket. It is a machine he can ride and drive at once; it is a vehicle he can keep in the house like an ordinary safety bicycle, and he can always get home on it should anything by chance go wrong. In a word, the motor bicycle will introduce the pleasures of moting to thousands of cyclists who would never otherwise be able to participate in the new pastime.[47]

---

[44] *Motorcycling*, 13 December 1909, cover.
[45] *The Motor*, 2 December 1903, ads. See also Armstrong, *Bouverie Street to Bowling Green Lane*, plate 22, in which artist George Moore first moots the sidecar as a solution for how a courting couple might be more intimate on the move (taken from his 'Love laughs at – Motor difficulties', *Motorcycling and Motoring*, 7 January 1903).
[46] *Motorcycling and Motoring*, 12 February 1902, p. 12. For an account of the starting up of the new motoring magazine, see Armstrong, *Bouverie Street to Bowling Green Lane*, chap. X.
[47] *Motorcycling and Motoring*, 12 February 1902, p. 12.

As with Dangerfield's previous magazine, *Cycling*, launched in 1891, his new magazine was rather irreverent and forward-looking. The earliest cycling magazine editors, according to Armstrong, had resisted change, even being scornful of developments such as the then new 'safety' bicycles and pneumatic tyres, and the cycling and motoring press tended to put its advertisers before its readers. For most of the earliest mainstream cycling and motoring magazines, critical analysis of a dire product usually extended no further than damning with faint praise. In that light, *Motorcycling and Motoring* (and after 1903, *The Motor*) could be rather shocking. The magazine's thoughts on the De Boisse Voiturette are a case in point. Regarding this French three-wheeler, with its engine mounted over the single front wheel, and the whole steered by tiller, *Motorcycling and Motoring* concluded, 'We must say that the machine is not at all pretty, the engine being in an awkward position, and we should think the passengers would enjoy rather more smell than they desired.'[48]

The circulation policy evidently worked. Costing one penny, and edited by former cyclists and magazine editors Edmund Dangerfield and Walter Groves, *The Motor* was hugely popular. Adopting a smaller font than *The Autocar*, it appeared every Tuesday, to give it a head start over, in particular, Thursday's *The Autocar*, on the previous weekend's news. Within two months the motor-cycling magazine was already featuring motor cars (Edge's 50hp Napier, capable of 80mph, with photos, was the subject of its first article on a motor car, which the magazine declared a 'pleasant change').[49] And, within the year the magazine had shifted its policy to support the new breed of light car, publishing a 'light car special' in late 1902,[50] and a 'Light cars' section thereafter. The magazine was now intended for the 'modest motorist',[51] and had relaunched itself as *The Motor incorporating Motorcycling and Motoring* (this latter bit was soon silently dropped). The magazine was now suggesting that it was difficult to say where the motor-cycle really ended and the light car began, but it thought a 'broad line can be drawn by dealing with all forms of light motor vehicles which appeal to those of moderate means, and this policy is to be our new departure'.[52] Its conversion to the cause of the light car within twelve months of launch as a motor-cycling magazine was indicative of a rapidly changing market and the editors' ensuring they kept sales up (and not least Dangerfield's intuition that 'motor-cycling was destined to make very little headway in the next few years').[53] Sales of *The Motor* were soon in the 30,000s, outselling all competition put together.[54] Its appeal shifted: we want to 'foster and

---

[48] *Motorcycling and Motoring*, 17 December 1902. For De Boisse, see G. N. Georgano (ed. in chief), *The Beaulieu Encyclopedia of the Automobile* (London: The Stationery Office, 3 vols, 2000), i, p. 395. See also Armstrong, *Bouverie Street to Bowling Green Lane*, p. 10.
[49] *Motorcycling and Motoring*, 26 March 1902, p. 108, for example, a run on the 'fastest car in the world'.
[50] Special light car edition, *Motorcycling and Motoring*, 3 December 1902, pp. 294–311, 313–14; carried over to 10 December 1902, pp. 323–6. See also 'light cars at the Paris Show', *Motorcycling and Motoring*, 17 December 1902, pp. 349–51.
[51] See Horner, 'The emergence of automobility in the United Kingdom', pp. 56–75. *The Motor* was ultimately absorbed by its rival, *The Autocar*, in 1988.
[52] *Motorcycling and Motoring*, 5 November 1902, p. 208.
[53] Armstrong, *Bouverie Street to Bowling Green Lane*, p. 70.
[54] *The Motor*, 19 July 1903, ads. More of *The Motor* was published than any other four competing magazines combined. By its 29 July 1903 issue, circulation was said to exceed all other magazines

THE AUTHOR AND HIS MODEST MOTOR.

Figure 5.5 Major C. G. Matson in the driving seat of his De Dion, about 1903. The young lad is his 'shuffer'. Image: Major C. G. Matson, *The Modest Man's Motor* (London: Lawrence and Bullen Ltd, 1903), frontispiece.

encourage the light vehicle movement', it said, but, wanting to have its cake and eat it, it reassured the manufacturer and user of the motor-cycle that it would continue to be interested in the 'evolution of the cheaper forms of light four-wheeled vehicles. Our programme will be "Motoring for Men of Moderate Means"'.[55] Reducing motor cars to a standardized form, in order to encourage the new motor traction as a trouble-free leisure device, had ready appeal for a new breed of consumer: the 'modest motorist', or variously, 'The man of moderate means'.

According to *The Motor*, this was the man 'in the majority'. He had been a cyclist, they said, and has now moved onto a motor-cycle. Now knowledgeable and informed, he yearned for company, so he bought a forecar. His friends copied him, but now he felt a yearning to grip the steering wheel. In short, he was now ready for a motor car. This 'departure', the magazine continued, had not depleted the motor-cycling fraternity because he had introduced so many to it – this was the mission of *The Motor*.[56]

Whether the 'modest motorist' was an invention of *The Motor* magazine is unclear, but other writers picked up on the phrase of the moment. N. G. Bacon did so for her article in the *Girl's Own Paper*. Major Matson, a contributor to the *Badminton* magazine

combined. It went from 21,000 in January 1903 to 36,000 in December: 3 February 1904, ads. Sales were then six times that of any other magazine.
[55] *Motorcycling and Motoring*, 21 January 1903, p. 450. See also Horner, 'The emergence of automobility in the United Kingdom'.
[56] *The Motor*, 10 February 1904, p. 3.

and *Daily Mail*, published his *The Modest Man's Motor* (1903), an entertaining account of running a car for not much more than ten pounds a year, in part by using solid tyres and therefore not needing expensive pneumatics with their propensity to blow out.[57] (*The Motor* was unimpressed with Matson's depiction of the modest motorist: the book was well spoken of, it conceded, but 'the title is not a felicitous one, since it hardly expresses the meaning of the motor for the man of moderate means'.)[58]

Running costs

In offering a definition of the typical 'man of moderate means', *The Motor* suggested this was probably a middle-class tradesman occupying the dwelling house above his business premises. Or he might be a professional man who lived in the suburbs and paid a rental of £50–100pa; or maybe a retired tradesman or official with a fixed income from his investments or pension.[59] But whereas *The Motor* tilted towards some measure of social inclusivity, the Automobile Club saw matters differently. In late 1902 Captain Kenneth Campbell (b. 1863) delivered a paper for the Club using the same language – 'Motors for men of moderate means' – but with a very different definition. At that meeting Major Lindsay Lloyd (1866–1940)[60] had defined a man of moderate means 'as one who rode third class in trains, used penny omnibuses, and walked on occasion'. It was just this kind of person who Campbell felt might get above his station: when he 'got on a car, he lorded it over others, and must expect to pay for the luxury. [. . .] The whole question was one of mileage, persons carried, and luxury. If they worked that out, they would find that a car was a very thing indeed'.[61]

The Automobile Club was, naturally, looking out for its own, and had a clear interest in ensuring the newcomer knew exactly how much motoring might cost him, whereas *The Motor* saw, instead, a huge potential market for motoring for those who might be persuaded if they thought how little it could cost. 'Cyclomot' in *Motorcycling and Motoring* commented on the ridiculousness of Campbell's talk, saying he knew people who had never owned a horse, and could not afford to run a pony or cart, but could afford to run a car.[62] Campbell's talk went on to form a chapter in subsequent editions of Harmsworth's Badminton book,[63] a book *Motorcycling and Motoring* had already condemned: 'Our contention that up to now the writers on motor subjects have written over the heads of the people who will be ultimately the motor trade's best

---

[57] Major C. G. Matson, *The Modest Man's Motor* (London: Lawrence and Bullen Ltd, 1903), pp. 1–4; N. G. Bacon, 'Modest motoring', *The Girl's Own Paper*, 5 December 1903, pp. 148–50.
[58] *The Motor*, 25 November 1903, p. 414.
[59] *The Motor*, 7 June 1904, pp. 481–3.
[60] See E. K. S. Rae, 'Unconventional portraits of leaders in motorism: Major F. Lindsay Lloyd', *Automotor Journal*, 25 September 1909, pp. 1151–3.
[61] *The Car Illustrated*, 28 January 1903, p. 326.
[62] *Motorcycling and Motoring*, 4 February 1903, p. 497.
[63] For example, in the fourth edition, Claude Johnson reworked the talk for his 'Motor cars for men of moderate means' which showed how costs could vary wildly, depending on tyres (whether changed), horsepower etc.: [Lord] Northcliffe (ed.), *Motors and Motor Driving* (London: Longmans, Green & Co, 4th ed., 1906), pp. 379–90.

customers is amply borne out by the critics of this book, who state that it is written for rich men.'[64] So, where Campbell was seeking to talk UP the cost of motoring, to dissuade the modest motorist and reassure his wealthy audience of their exclusivity, *The Motor* was championing just the opposite; this stance can be seen in its spat with the Automobile Club when the magazine attempted to put on its own trial (discussed in Chapter 4). This led to a barrage of (conflicting) articles on just how much it cost to run a motor car. Campbell's talk included his meticulous costings – the car, first of all, for £525 (and therefore well above light-car prices), and eye-watering running costs of £205 per annum. This excluded the cost of his groom (£130 per annum, not forgetting ten shillings for his Christmas present), and he assumed the motor car would be kept in a coach house. With its readership in mind, *Motorcycling and Motoring* found the talk a 'disappointment', since Campbell's definition of 'moderate means' was an income of £2,000–2,500.[65] Instead, *The Motor* found the term 'moderate means' quite elastic – it could mean having an income of anything between £250 to £750.[66] Working on the basis that anyone who could afford a motor car was sure to want to buy one, *Motorcycling and Motoring* pointed out that 'it is clear that the market [for the light car] is not as unlimited as it is apparently thought to be'.[67]

Motorcycling and Motoring [soon, *The Motor*] also recognized a meaner day-to-day reality for its modest motorist: 'These folk rarely possess accommodation for anything larger than a bicycle' – no coach house for them – and their first consideration was storage; it was considered essential at the time to store a vehicle covered up overnight. The necessity of obtaining admittance at any hour of the day or night, including Sundays, precluded storage in the high-street garage or local repair shop. Where such accommodation was available, charges were often high and it was impossible to get in or out at abnormal times. Garages, that is, business premises, it continued, were often no good because the car owner would get the blame for tools that went missing. Private storage was better, it said, but more expensive. A study of Cheshire vehicle registrations to 1914 bears this latter scenario out. Starting in 1904, these records show that a motor car or motor-cycle, probably brand new, was registered at a Cheshire address; this would often be somewhere comfortably middle class (The Firs, The Vicarage etc.) in one of the wealthier outlying towns (Lymm, Bowden, etc.). The records show how such vehicles exchanged hands many times over just a few years, ending up at a working-class address in a suburb of, say, Manchester or Salford.[68] It is these owners, with restricted or no access for a vehicle on their property, that *The Motor* was referencing.

[64] *Motorcycling and Motoring*, 30 April 1902, p. 206. Cf *Motoring Illustrated*, at 3d, pandering to the more socially select reader, and which praised the 'masterly manner' of its 'experts': 'the volume should have its place on the library shelves of all lovers of the automobile' (3 May 1902, p. 217).
[65] *Motorcycling and Motoring*, 28 January 1903, pp. 482–3.
[66] *The Motor*, 7 June 1904, pp. 481–3. The average weekly income for a 21-year-old male in 1906 was 28s (the 10th, 25th, 50th, 75th, 90th percentiles were 17s, 22s, 28s, 35s, 43s): table 6.2, from Ian Gazeley, 'Income and living standards, 1870–2010', in Roderick Floud, Jane Humphries and Paul Johnson (eds), *The Cambridge Economic History of Modern Britain, Vol. 2: 1870 to the Present* (Cambridge: Cambridge University Press, 2014), p. 155. The threshold for income tax was £160.
[67] The magazine used *Whitaker's Almanac* to extract the numbers of people in income brackets: 340,425 had a gross income of up to £1,000, of whom 210,879 were up to £200: *Motorcycling and Motoring*, 3 December 1902, p. 294.
[68] Horner, *The Cheshire Motor Vehicle Registrations, Vol. 1*.

Whatever, and in sharp contrast to Campbell, *The Motor*'s modest motorist need only allow ten pounds a year outlay. To keep to this cost, *The Motor* said, the owner should patronize the local repair shop and look after the car himself. It was also understood that the owner who had been 'in business' all day would want to spend his evenings with his family, and would not feel in the mood to spend an hour on 'adjustments'; this neglect, though, would mean bigger repair bills. It suggested that the lowest running expense for a two-seated car was 3d per mile, not allowing for tyres, repair work or depreciation. Here, it thought, the tricar scored all round, costing £85 to buy with subsequent running costs of 1½–1¾d per mile – storage would cost the same, but depreciation would be less.[69]

Writing elsewhere at that time, and with a wealthier, clubbable audience in mind, Filson Young was of the mind that anyone who could not readily produce fifty pounds for eventualities would be unable to afford to go motoring.[70] *Motoring Illustrated* in their 'Motoring for the Million, not the Millionaire' article in 1902, published details provided by eighteen motorists on the costs of running a car, which varied enormously (£5 to £500 per annum). These figures were either 'valueless or most instructive', and show how the car was 'managed or mismanaged', which would determine whether or not it was an 'inexpensive and delightful amusement'.[71] In trying to make sense of this variation, in *The Motor*'s 'Opinion' column, the writer suggested that the potential buyer would already know whether he could afford to buy a motor car or not; the issue was the cost of the upkeep. Take the family who would not have kept a horse, it explained. A car would cost up to £200, and then £20 upkeep for 4,000 miles a year. Then fifteen pounds for tyres and two guineas for a licence, let's say £42 2s as ample. But the magazine thought even that was excessive, and suggested an annual cost of £27 10s, plus the cost of stabling (garaging), and here the owner was advised to build his own garage to keep costs down. Depreciation should not exceed £25 per annum.[72] Much less, then, than owning a horse.

Readers responded to this with their own costings, but all in broad agreement: 'J. A.' had bought his car privately the previous August (1902) for £75 – a 5hp 1-cylinder model, make unspecified. He had become self-reliant and engaged the local wheelwright only when machine work was necessary.[73] Meanwhile a report on the 5,000-mile reliability run of a 12hp Siddeley had revealed a total running cost of £40 10s. In its 'Opinion' piece on the 'The cost of motoring', it was pointed out that the actual cost would have been only £19 15s were it not for the two tyres necessary – according to Bolster, one tyre could cost more than a year's petrol[74] – and that of the thirty-three days taken for the exercise, only three were dry (therefore wear and tear was above the normal). With an average mileage of 2,500, *The Autocar* deduced an average annual cost to the

---

[69] *The Motor*, 7 June 1904, pp. 481–3.
[70] Filson Young, *The Happy Motorist*, p. 48.
[71] *Motoring Illustrated*, 24 May 1902, p. 300.
[72] *The Motor*, 8 April 1903, p. 194.
[73] *The Motor*, 15 April 1903, p. 209.
[74] John Bolster, *Motoring Is My Business* (London: Autosport, 1958), p. 14.

motorist of twenty pounds.[75] It was spun not so much as, can you afford to buy and use a car, as can you afford not to; Ivan B. Hart-Davies, an insurance broker, said on using a motor-cycle for business, 'my income would be halved if I went back to trains or ordinary bikes'.[76]

Reading *The Motor*, the consumer would also be tempted to believe that, while he could start motoring for as little as £100 for the motor vehicle – that would buy a 'reliable' two-seated petrol car – he might like to consider stretching to, say, £250, which would secure something 'more refined'. In addition, that new motorist would also need to set aside at least ten pounds for tools and extras[77] – increasingly, there was a need to 'equip' the car, as after-market accessories became more desirable. 'Nemo' pointed out how when he bought his light two-seater, he found himself paying an extra fifty pounds, to include a horn, spare wheel, speedo, clock, lamps, bulbs, batteries, hood and windscreen. *The Motor* had its part in playing up these 'essentials', one editorial pointing out that these should not be 'skimped on'.[78] Levitt urged her lady readers to buy as extras a hood (£18–20); folding windscreen (£10); front lamps (£6) and rear lamps (from £1). A waterproof rug (£1–2) and a folding third seat (£15) were also recommended.[79] Meanwhile, motoring magazines carried full-page adverts for retailers such as Dunhill's or Gamage's who sold such items.

Strategies for buying

The new modest motorist had several options if he could not afford to buy brand new. In its article 'Co-Partnership Motoring: Some suggestions as to how the cost of motoring can be divided' in 1913, *The Motor* suggested that while a wise motorist was 'very jealous about his car', nowadays 'motoring is no longer an art confined to a small number of expert people'. It was no longer a 'fearfully complex and delicate piece of mechanism which dare not be trusted out of one's hands'.[80] In other words, shared ownership amongst the non-technical was now a real possibility. Localized motor agencies sprang up rapidly, and were likely to handle second-hand vehicles; the Cheshire records show a large number of motor cars and motor-cycles passing through the hands of agents, presumably to be sold on. Most local agents probably sold second-hand, but there were exceptions such as Wauchope's, whose adverts in *Motorcycling* offered a range of some 200 motor-cycles new and used for sale; buying used started at £8 10s.[81] Plus, *The Motor* offered a 'Deposit System', whereby a deposit could be lodged with the magazine while a vehicle advertised therein was inspected,

---

[75] 'The cost of motoring', *The Autocar*, 14 February 1905, pp. 37, 39. See also Anders Clausager, *Wolseley*, p. 76.
[76] *Motorcycling*, 5 August 1910, p. 329.
[77] *The Motor*, 14 November 1905, pp. 355–6.
[78] *The Motor*, 22 March 1910, pp. 260, 282.
[79] Levitt, *The Woman and Her Car*, p. 18.
[80] 'Co-partnership motoring', *The Motor*, 11 March 1913, p. 272.
[81] *Motorcycling*, 20 December 1909, supplement, p. i.

and that deposit would either be passed on to the seller, or returned to the buyer, all for a 1¼ per cent fee.[82]

The letter-writing readers were under no illusion that the purchase of any motor remained fraught. 'Auto' wrote that buying second-hand was more difficult than buying a horse. With a horse, he pointed out, the 'usual guarantee' was to have the horse for a week, have a vet's exam, and if unsound, the seller would have it back. This cannot be done with cars, but why not try? he said. Asking this provoked a reaction in the letters column – for example, L. Savory, probably a salesman himself, pointed out how unfair this was to the seller, since an ignorant purchaser could cause much damage to a vehicle in a week.[83] It was well understood that it would be easy to rip off many customers. Matson said, 'buying a horse is a ticklish business all the world over, but it is child's play compared to buying even a new motor-car [. . .] and when it is not new the difficulty is far greater'; by this he meant, with good reason, that it might be a 'freak' or 'experimental',[84] as cars such as the Mabley or Carpeviam would turn out to be seen. Letter writers also corresponded with each other, exchanging views and information.[85]

Acknowledging that there were very few motor cars, new or second-hand, in the earliest days, I contend that buying second-hand was the way that most people took their first step into motoring. *Motorcycling and Motoring* from the outset endorsed this method, showing that prices could vary from £20 to £35 for many of what it thought were then standard and well known makes such as Excelsior, Werner or Quadrant.[86] Once the magazine had become *The Motor*, it continued to run many features on buying second-hand; in one it assured readers there is 'a lot of very good fun in it', that the product will have been tried and tested, that the parts that are going to wear will have done so; and it will be available for a moderate price.[87] Indeed, in recognition that buying second-hand was so popular, *The Motor* and others ran features on how to be confident when buying second-hand. When buying a used light car, 'Magneto' advised a minimum fifty-mile run; a test (stopping, holding still and restarting) on a hill of at least a one-in-ten gradient; and checking the action of the clutch and the condition of the tyres.[88]

Buying second-hand was accomplished in many ways. Vehicles were advertised in the community by local newspaper adverts, both for private sales and from local agents. But many sales were undertaken by methods which have not left any record, for example, by word of mouth, from a neighbour, or between workmates; evidence for this is inferred from the Cheshire records by subsequent owners of vehicles having the same occupation, or living on the same street. To what extent the reams of practical advice offered by motoring magazines for buying second-hand was adhered to is unknown, but it is likely that most paid little heed. 'Automan''s advice for buying

---

[82] *The Motor*, 27 January 1904, ads.
[83] *The Motor*, 11 February 1905, p. 202; 25 February 1905, p. 302.
[84] Matson, *The Modest Man's Motor*, p. 12.
[85] See John Hope's experience with a deficient Benz, where he confided he had had lots of private correspondence: *The Autocar*, 13 January 1900, p. 37.
[86] 'Hints and wrinkles', *Motorcycling and Motoring*, 23 April 1902, p. 176.
[87] *The Motor*, 7 June 1904, p. 486.
[88] *The Motor*, 18 April 1905, p. 287.

second-hand was to avoid a car by a company no longer manufacturing because of the difficulty with spares; to get independent advice; and to take a reasonable economic view of a car whose owner had 'bought larger'.[89] But most will have been taken in by the smooth sales patter of the salesman; it has already been shown how the Williamson writers were sold a 'dud', while *Punch* picked up on this phenomenon – for example, in 1902, a cartoon showed the rider blurred due to the excessive vibration of his motor-cycle, with the caption, 'How Jones felt on a second-hand motor bicycle of the vibrating kind which he had bought for a "mere song"'.[90]

O'Connell has pointed out that for the interwar period, buying by instalments (on 'terms', or 'tick') was an option, even if a little socially embarrassing to do so.[91] There is much evidence, though, to suggest that this method was widely available in the period to 1914. Bicycles had long been available by these means: Swift cycles, for example, selling for ten to twenty guineas, were available in 1900 on the 'gradual payment system'.[92] *Motorcycling* pointed out, in 1910, that should the initial outlay be too much for a new starter's pocket, 'there are several firms manufacturing the very highest grade machines who undertake orders on the gradual payment system'.[93] Agencies such as Wauchope's of London carried regular half-page adverts, selling second-hand motor-cycles for cash or instalments.[94] The Civil Service Motor & Cycle Agency Ltd of London placed a half-page advertisement of second-hand cars (starting at £55 cash) and motor-cycles (from £26 in cash, or by instalments).[95] The Nymph motor-cycle was available on 'Extended payments on easy terms',[96] while the Wilbee motor-cycle could be had for 'cash, instalment or exchange'.[97] Second-hand cars were also available on an exchange basis – the small ads in *The Motor* in 1903 enabled the purchase of larger cars second-hand, for example, a 6½hp Darracq, in good condition for £120, and where a good motor-cycle would be taken as part-payment.[98] The small ads in local newspapers (often still under the heading 'Cycles') were used to sell used cars locally. It is unclear how many customers took advantage of payment on terms, or as an exchange, but by 1913, *The Cyclecar* was reminding its readers that 'Several firms of good standing' will sell on instalments to suit the buyer, which suggested that this remained a popular method.[99]

Motor cars could be bought new on terms. The American Crestmobile, already a cheap car at 100 guineas (£105), could be secured for a £29 down-payment and twelve instalments of £7 5s (totalling £116). This 'Light touring car' was the 'horseless carriage for the novice. The only car that can be really driven and kept in repair by persons with no mechanical knowledge'. Its image changed rapidly, from a typically American

[89] *The Autocar*, 18 April 1905, p. 289.
[90] *Punch*, 3 September 1902, p. 158.
[91] O'Connell, *The Car in British Society*, pp. 25–32.
[92] *Bicycling News and Motor Car Chronicle*, 2 May 1900, p. 19, ads.
[93] 'The businessman and the motor-bicycle', *Motorcycling*, 28 June 1910.
[94] For example, *Motorcycling*, 20 December 1909, supplement, p. i.
[95] *The Motor*, 26 August 1903, ads.
[96] *The Motor*, 15 July 1903, ads.
[97] *The Motor*, 12 August 1903, ads.
[98] *The Motor*, 15 July 1903, ads.
[99] 'Counting the cost of a cyclecar', *The Cyclecar*, 7 May 1913, pp. 617–8.

flat-bed buggy to something more akin to a light car. It promised readers, 'We are daily supplying these cars to COUNTRY GENTLEMEN – ARMY OFFICERS – DOCTORS – CLERGYMEN and PROFESSIONAL and COMMERCIAL MEN, who have had no previous experience of motors, and who are driving and looking after them without the assistance of chauffeurs.'[100]

This appeal to particular trades of high standing was nothing new and was made with medical doctors especially in mind. Mom has suggested that the particular place of the 'doctor's car' has been overblown,[101] but certainly, magazines occasionally published 'doctor's car' special editions; *The Autocar* had its first in 1899.[102] In an article, 'Motors for medical men', Dr Hills of Belvedere in Kent wrote to *Motoring Illustrated* in 1902 to describe how he had dispensed with three horses, a trap and a carriage, all for a Daimler motor-car, which was now costing him about a penny a mile. The conclusion was that there can be 'no doubt' that over the coming years medical men will use more motors.[103] As with all occupations though, the 'doctor' in society covered a potentially wide range of income, expectation and experience – Dr Tracey, for example, was the first in his village, in 1907, to own a motor-car and was, despite its clear faults, sufficiently smitten to buy another of the same make.[104]

Others were still waiting. J. R. Ratliffe's letter in *Motorcycling and Motoring*, in 1902, said that as a 'medical man' from Birmingham, and 'like many others', he was 'waiting for the time when there shall be a really suitable car for us'. He thought that the new light car, being paraded by the magazine, had 'a great future before it for doctors'. However, of particular note is that Ratliffe was identifying his requirements in a light car. It should be a two-seater, simple, reliable, started from the seat 'for if you are going from one street to another only, you cannot be hurdy-gurdying in a frock-coat and top-hat every time and it is a waste of time to let the man do it'. He also named the price: 'It is not every struggling medical man who can afford three or four hundred guineas; eighty to a hundred pounds would be a popular figure (if I am not asking too much for my money).'[105]

Ratliffe was one of many who wrote in to identify what they wanted and how much they would pay for it. While all of them did not necessarily agree on these details, it showed how the consumer was feeding into the development of motor traction. *Motorcycling and Motoring* magazine played a part in shaping expectations for the nascent motorist and reporting their demands, usually as letters to the editor. On the one hand, it idealized the light car for the 'modest motorist' by, for example, placing full-size hand-drawn pictures on its front covers depicting a life of leisure and ease for the middle-class couple, facilitated by the motor car. Those on motor-cycles were encouraged to try a sidecar. They in turn were encouraged to move to a tricar, and

[100] *The Motor*, 26 August 1903, ads; 26 July 1903, ads; 8 July 1903, ads.
[101] Mom, *Atlantic Automobilism*, pp. 71–2.
[102] *The Autocar*, 7 October 1899.
[103] *Motoring Illustrated*, 3 May 1902, p. 214.
[104] Dr Tracey's indifferent experience with two Peugeots is useful for showing just how unreliable cars still were in 1908: Tracey, *Father's First Car*. See also Horner, 'The emergence of automobility in the United Kingdom', p. 68.
[105] *Motorcycling and Motoring*, 3 December 1902, p. 311. Being able to start a motor car from the seat remained unusual until the advent of the electric starter, first introduced in 1912.

again to a light car. But in feeding back to the agents and entrepreneurs, it also reported that

> The enquiries *which have reached us* [my italics] suggest that what is required is a car to accommodate two people, and to have room for all necessary tools and a certain amount of luggage. The motor mechanism must be reliable, must be as simple as possible, must be accessible, and must be able to take the car and its full load of passengers up any hill. Speed is not asked for, but our own experience leads us to believe that a turn of speed on the high gear will not be at all unacceptable even by those who profess to despise it. [. . .] The body should be roomy, with storage room under the seat and behind.[106]

The rise of the consumer

The magazines had then helped create and inform a new class of motor-traction owners, and while the gullibility of some readers persisted, fantastic sales pitches, such as Pennington's ravine-leaping motor-cycle, were rebuffed by an increasingly well-informed consumer. The new upmarket magazine *Motoring Illustrated*, in its first number, claimed to have identified motoring as a realm for consumption in distinction to magazines hitherto, which had been mere 'trade catalogues', 'read to some extent by motor car dealers'. Instead, it said, we will appeal to '"consumers" of motor cars – the ladies and gentlemen who own self-propelled vehicles, and ride in them for pleasure'. It continued, 'we dedicate *Motoring Illustrated* to the latest and best of English sports. We have no interests to serve except those of the motorist, no trade affiliations of any sort, no favourite brand of motor-car to "boom", no prejudices, and not a single enmity to which to minister'.[107]

The new consumer was increasingly identified as distinct from the previous world of Jarrott's 'old brigade' who increasingly was being recast as a relic of his time. By 1914, motorists thought of themselves as 'old timers' if they had been driving ten years before. The environment had changed so much that Henry Sturmey felt it necessary to suggest 'duties' by which the new motorist should abide, such as showing consideration for cyclists and passing horses with care. This was indicative of the new type of motorist: 'most people who drive cars have been [cyclists]', but 'there are drivers who have never cycled',[108] he said. 'Automan' asked what had become of our 'friends of the early days? Where are the Bollee's, Benz's and such like, or the belt-driven cars?',[109] referring to vehicles from the 1890s, few of which were ever available even then. In a review of a book, *The Motor* wrote that to be a good driver in the old days was to understand the engine and transmission, and the early motorist was a 'rare bird'; that was altered now.

---

[106] 'Light cars: Some of the requirements', *Motorcycling and Motoring*, 31 December 1902, p. 390.
[107] *Motoring Illustrated*, 8 March 1902, p. 1. For a discussion of the rise of motor-car culture see James J. Flink, *The Automobile Age* (Cambridge, MA: MIT Press, 1988); and more recently, Mom, *Atlantic Automobilism*, chap. 1.
[108] 'The duty of the motorist: A straight talk to new motorists', *The Motor*, 6 January 1914, pp. 1154–6.
[109] *The Motor*, 19 December 1905, p. 549.

The 'Old Brigade' and the New 'Steady and Careful Artisan'   111

Figure 5.6 The financier Edward Pennington made ludicrous claims in his catalogues, playing on wide public ignorance of the abilities of the new motor traction. Here, in G. N. Martin's illustration, the 'Flying Pennington Cycle' (1896) is made to leap a ravine of sixty-five feet: From H. O. Duncan, *The world on wheels, vol. 2* (Paris: privately pub., 1926), p. 688.

The roads are now 'full of swiftly moving traffic' – 'an entirely new set of conditions has arisen which needs special attention'.[110]

The status of the 'old brigade' was cemented with the setting up of 'old timer' networks, particularly after the First World War. Most notable was the Circle of Nineteenth-Century Motorists, formed in 1927 and open to anyone who could prove they had driven a motor vehicle prior to the Thousand Mile Trial of 1900. About 250 members were elected and participated in grand annual dinners, with, it seems, eyebrow-raising entertainment,[111] before declining membership caused it to be wound up in 1960. That it was necessary to have two references, and that each application was scrutinized by the honorary secretary Montague Grahame-White is an indication of the social credit that membership brought.[112] Edge was quick to join, and his handwritten notes on many of the application forms shows he was involved closely in

---

[110] Review of *How to Drive a Motor Car* (1s 9d): *The Motor*, 9 June 1914, pp. 888–9; and for 2nd ed., 21 July 1914, p. 1223.

[111] Their 1935 entertainment at the Trocadero Restaurant on Piccadilly Circus included an after-dinner cabaret presented by 'adagio dancers' Alexis and Dorrano 'from the Casinos on the French Riviera': letter, Montague Grahame-White to R.W. Buttemer, 15 October 1935, RAC Archive, ACQ4/6. The British Pathé clip of the dancers from the film *Danse Apache* (1934) is notably violent and shocking: see https://www.youtube.com/watch?v=I0DVt3xbecg, accessed 18 October 2019.

[112] See Grahame-White, *At the Wheel Ashore and Afloat*, chap. 40; Michael Edwards, 'The Circle of Nineteenth-Century Motorists', *The Automobile*, 23:10 (2005), pp. 48–50. See also RAC archive, ACQ2/1.

the vetting process thereafter. Fellow members included Henry Ford (1863–1947) and Herbert Austin. Open only to men (even guests at the dinners had to be male), this was truly a list of the clubbable. Edge was also in the Fellowship of Old-Time Cyclists, a fraternity of older cyclists who also enjoyed meeting over dinners. Membership was restricted to those born before 1873[113] and included many who straddled the cycling and motoring worlds.

Buyers of (new) motor cars were becoming more conscious of style, a prejudice encouraged by entrepreneurs who needed to sell subsequent motor cars to the same customers. Older cars were now referred to as 'crocks', compared with 'this year's model', and with a clearer expectation of how a motor car or motor-cycle should 'look', the 'crock' did not hold its value.[114] 'Anastasius' wrote to say that other motorists tell him he was wrong to want to keep a car for six years or more because 'cars become so quickly old-fashioned'. He thought the motor now stood where the 'push bicycle' did about fifteen to twenty years before. Between 1890 and 1895 the bicycle underwent rapid development, as did the motor car between 1900 and 1905. Since then, 'the main lines of motor car development have remained fixed'.[115] 'Anastasius' had a point.

Club membership for most, though, and for much of our period, was the privilege of the few and presented one barrier for widening participation in the motor-traction movement. While the Automobile Club wrestled over what sort of club it was, and wanted to be, that is, for members, for amateurs, and not for the trade,[116] the majority of new motorists – those who did not join any club – remained a submerged group, and details are known about them now only where, say, county vehicle-registration records survive. In Cheshire, for example, it is possible to trace a motor-cycle and its registered owners, bought, say, for sixty pounds new in 1904, which six owners later (and probably bought for a fiver) was at an address in a working-class terrace, with its owner a manual worker.[117] These later owners were the ones least likely to seek or benefit from club membership.

The Motor magazine had deliberately positioned itself in the market as a friend of the 'modest motorist', and held a scathing view of the Automobile Club's abilities to represent the same. Letter writers to the motoring press occasionally asked whether clubs served any purpose at all. In *The Motor*, 'Anti-humbug', for example, in 1910, pondered the point of motoring clubs. He said he belonged to, or was affiliated to, the

---

[113] See Derek Roberts, 'A short history of the Fellowship of Old Time Cyclists', *The Boneshaker*, 198 (2015), pp. 50–1.
[114] The small ads columns of local newspapers are full of second-hand vehicles for sale, but as an indication of how motor cars lost value even in the earliest days, T. R. B. Elliott recalled paying £250 in Paris for his first car, and then selling it a year later for £112: Elliot in Montagu, *The First Ten Years of Automobilism*, p. 20.
[115] *The Autocar*, 4 July 1914, p. 44.
[116] When the Automobile Club was being formed, for Simms, it needed to be 'essentially a members' club', but Salomons feared Simms simply wanted to create 'another disguised promotional organisation'. This wrangling is described in Brendon, *The Motoring Century*, pp. 32, 34.
[117] For Cheshire, examples of motor cars or motor-cycles changing hands three or four times within a few years are frequent, but six or more times is not unusual. See, for example, M 457, a 2¾hp Ariel motor-cycle registered, presumably new, on 25 April 1904 to an address in Crewe. The motor-cycle is on its sixth owner by 11 June 1909, a fitter at a colliery, living in a terraced property in Tunstall, Staffs: Horner, *The Cheshire Motor Vehicle Registrations, Vol. 1*, p. 155.

AA, RAC and Motor Union (MU), via a local club, 'but have never obtained any benefit direct or indirect from the RAC or MU, nor have I encountered anyone who has . . . The RAC is conducted with a pomp and solemnity suggestive of the ancient Mexican priesthood, and tells us now and again that this or that car ran a thousand miles or so, but what has it ever done for the average motorist?'[118]

'Anti-humbug', then, is likely to be representative of a widening mass of motorists and motorcyclists, and his (or her) case reminds us that most motorists, motorcyclists and cyclists did not belong to any club. This was almost certainly down to choice (rather than being unable to afford annual subscriptions), but meant the social capital, and access to information and assistance that club membership brought was increasingly irrelevant to the needs of the cyclist or motorist. (Some) clubs soon broadened their social composition and character. For some cycle clubs this had happened by the late 1880s when they offered 'advantageous terms' allowing the purchase of a bicycle by weekly payments. Shops started to offer hire-purchase terms to club members, and a glut of cheap second-hand 'ordinaries' came onto the market from the late 1880s. Some clubs became less militaristic, their parades assuming a more festive air, and they were prepared to hire out country cottages for use by members.[119] The standard dress of the ABC – grey cloth suit and black stockings – was abolished in 1891. The uniform of the Black Anfielders, a cycling club formed in 1879, black from head to foot, all 'decent, sober and correct', was not actually abolished, but 'just faded out'.[120] The Manchester Motor Club founded in 1904 (distinct from the MAC, and having a broader social appeal, having started off as a motor-cycle club), offered lower annual subscriptions of 10s 6d to draw in motorcyclists, and of twenty-one shillings for car drivers. It had a more 'open' programme than its counterpart, to increase its popularity and membership (then standing at 130). This included hill-climbs for motor-cycles; events for non-trade participants; tricar hill-climbs open to all members; car owners' hill-climbs for non-trade; a reliability trial; a speed judgement test; and a hill-climb for car owners within a certain radius of Manchester.[121] These were all clear signs of a more socially inclusive approach.

One way in which the consumer could identify himself as a force resistant to the pressures by clubs such as the Automobile Club was through attitudes to cleaning motor vehicles. There had been a prevalent association of cleanliness with social class, which had meant horses, owned by the wealthy, were often well groomed, and carriages easy to clean. The ready availability, and take-up, of the new modes of road traction was changing all that, and challenging long-understood 'duties' for staff in service. *Punch* featured a cartoon in 1896 portraying two young ladies cleaning their own bicycles in their grand hall, their aprons oily, as snooty footmen look on. 'It is not the business of ducal footmen to clean the family bicycles', the caption read. 'The ladies Ermyntrude and Adelgitha have to do it themselves'.[122] It has been shown that the Automobile Club at the time of the Thousand Mile Trial in 1900 fretted over the potential spectacle of

---

[118] *The Motor*, 5 April 1910, p. 378.
[119] Lawson, 'Wheels within wheels', pp. 132–4, 137–8.
[120] [anon.], *The Black Anfielders*, p. 2.
[121] *The Motor*, 11 April 1905, p. 269.
[122] *Punch*, 4 January 1896, p. 6.

road-soiled motor cars driving into towns. The Club urged drivers to wash (or rather, have washed for them) their cars before they were presented to a wider public; being seen in a clean motor car was vital for the credibility and acceptance of the new road traction. This subject waxed and waned in the letters columns, but the need to be seen in a clean vehicle ceased to be universal, particularly as the motor-cycle and motor car were adopted by a wider consumer base. 'Owner-driver' wrote in to say that a man told him he gave up motoring because of the bother of washing his car after every run, while 'O-d' said that his car had been washed for the first time since October: 'For the owner-driver to waste his time over washing his car seems to me utterly ridiculous'.[123] The message from the magazines and clubs to keep a vehicle clean became increasingly irrelevant with increasingly wide ownership.

Instead, convenience – and changing motor-car design which increasingly hid away oily mechanical bits – defined the new consumer. *Motorcycling and Motoring* magazine recognized this trend:

> The chief object aimed at in the light car is simplicity and convenience; reliability, of course, being understood. It must be a car that the merest amateur can mount and drive without giving himself much unnecessary trouble. The seating arrangements, moreover, must not depart greatly (at first) from an ordinary carriage; especially must there be no bother in getting in and out either for the driver or his companion. And, as it is possible that this companion may sometimes be a lady, there must be no mechanism, or even lubricating gear, which can soil her apparel.[124]

(Dorothy Levitt always carried a cover-all apron to wear for any repairs or maintenance.)

Motor cars were increasingly marketed for their simplicity – the starting from the seat, to avoid turning the starter handle, the 'bolt-on' ('Stepney') spare wheel for the punctured tyre, and so on. Underpinning this new and modest-cost market was a constant looking forward, and a rejection of what had gone before. Writing in 1909, *The Motor* claimed 'it was impossible [five to six years ago] to buy for, say, £150, a car that today would be considered to be worth having at all. In fact, the cars at anywhere near that price were once, in our hearing, described as "horrid things", and the description truthfully fitted them'. 'We do not say that progress is at an end, nor that improvements are not possible', but they did say that 'no man who has hesitated need hesitate longer'.[125]

The motor journalist 'Pilgrim' suggested that cycling had also once been accompanied by a great deal of 'show': 'those who did not appear smartly attired and with smart bang-up-to-date machines, and who could not afford 9s or 10s a day for touring being made somehow to feel very uncomfortable'. But now cycling is 'everybody's mount. Smartness now has nothing "classy" about it; it merely counts as a personal fancy, and one can tour in any garb one likes, on a machine of any age, and patronise any

---

[123] *The Autocar*, 16 April 1910, p. 523.
[124] 'Comfort in the light car', *Motorcycling and Motoring*, 31 December 1902, p. 389.
[125] 'A new motoring era', *The Motor*, 2 November 1909, pp. 425–9.

class of accommodation without feeling in any way "out of it".[126] This confirms the 'dark ages' of cycling between 1900 and 1920, described by Oddy, when there was a 'perceived declining cultural status' of cycling coinciding with a 'technological and aesthetic "closure"', although Oddy shows that the story is not quite that simple.[127] This also suggests a 'de-clubbing' by a broad swathe of new cyclists from the mid-1890s, a practice slowly filtering through to motor traction a decade or so later.

'Pilgrim' was right that it really could cost ten shillings or so for an overnight tour by motor-cycle. He offered as evidence typical costs for 'classy' hotels: lunch 2s 6d, bedrooms 3s 6d or more, breakfasts from two shillings, and a shilling for 'attendance' to be paid as a tip again and again. On the other hand, he reckoned, nobody need pay these costs. An out-of-town place he knew charged him three shillings for his evening meal, bed and a ham-and-egg breakfast (less than half what he would have paid in the town). And for the midday meal, he knew of a pastry-cook who sold him roast duck for 1s 3d, or half what he would have paid at a hotel. And cheaper still, there was the picnic; it could be a bit tame for one, but 'rather good fun' with more. This way, touring 'in some such modest way', in an 'average district', could be kept down to 7s 6d or eight shillings a day, including petrol and oil.[128] The 'new steady and careful artisan' was starting to find it was not necessary to spend so much money out on the road.

Women

Women as consumers of the early motor traction tend to have little mention. The Automobile Club did not admit them as members, nor could they apply to local clubs such as the Manchester Automobile Club. Bicycle clubs, even cycling magazines, had frowned on issues such as 'rational' dress, although the arrival of the 'safety' bicycle permitted an influx of women cyclists, just as the increased standardization of the motor car did for female motorists.[129] But there had always been pioneering female cyclists and motorists, although they tend to be recorded only when they had done something exceptional. Our glimpses otherwise are tantalizing.[130] The Anfield Bicycle Club, formed in 1879, initially conducted 'quiet cycling', where 'even ladies toured, on tricycles of fabulous weight and shape'.[131] Smith recalled that there were lady members in the ABC; two are identified in the 1890s, one being Eleanor Rose Sharp (1872-1914) who became the first Mrs Edge, and the other, Mrs Moore, although 'we never saw

---

[126] 'The Pilgrim', 'The "swagger" motorcyclist', *Motorcycling*, 30 August 1910, pp. 387-8.
[127] Oddy, 'Cycling's Dark Age?', pp. 79-86.
[128] *Motorcycling*, 30 August 1910, p. 388.
[129] See Bury and Hillier, *Cycling*, chap. 8, 'Cycling for ladies', which also observed that prior to the 'safety', 'lady cyclists' would never have been expected to use a 'bicycle' (i.e. 'ordinary'), at p. 250.
[130] For literature on gender and mobility, see for example, Clarsen, *Eat My Dust*; Scharff, *Taking the Wheel*; Virginia Scharff, 'Gender, electricity and automobility', in Martin Wachs and Margaret Crawford (eds), *The Car and the City: The Automobile, the Built Environment and Daily Urban Life* (Ann Arbor, MI: University of Michigan Press, 1992); Wintle, 'Horses, bikes and automobiles', pp. 66-78; Wosk, *Women and the Machine*; Jungnickel, *Bikes and Bloomers*.
[131] [anon.], *The Black Anfielders*, pp. 2-3.

[her] cycle'.[132] Eleanor Sharp was a frequent rider, and not always with Edge, out in all weathers on club runs; she was interviewed for *The Lady Cyclist* in 1896 where she confirmed she also wore 'rationals', because otherwise 'the skirt flaps about so terribly when riding quickly'.[133] Indeed, there were probably other women active in the ABC at that time: a chance entry by Smith for 3 December 1893 records how the 'Ladies [were] scorched horribly', presumably by passing cyclists not part of the ABC. Ladies, though, were not invited to the annual club dinners and festivities; they were tolerated in the Automobile Club premises, but, as one member put it, they would 'demolish any barrier of rules which the mere man may erect to guard some private retreats for himself'.[134]

This hostility to women underscores the problem in establishing the exact role of women in the early cycling and motor traction movements: their enforced invisibility. Where male editors, entrepreneurs and photographers have made them visible, women are often restricted to being 'placed' in motor cars for illustrative purposes – for the covers of magazines, or for copy within. While this demonstrates that women played an active role as consumers, often such placement was intended to denote respectability and gentility for the sport, and ease of use of the motor vehicle. *The Cyclecar* magazine, for example, considered of the cyclecar genre, 'A lady would have no difficulty handling the machine'.[135] The publicity photograph of the new Pegasus 6hp light car featured in *The Motor* had two young women aboard, implying an ease with which the car could be driven.[136] In its gossip pages, *The Cyclecar* wondered, 'Should ladies drive motor vehicles? When it comes to a machine so easy to handle as the simplest type of cyclecar, we say why not? What do our readers think?'[137] Catering was so often the exclusive job of the female partner: 'Cyclecarist's Wife', for example, describing their trips away, said they stopped daily by the wayside carrying a tea and a luncheon basket (prepared by her). Overnight they wanted a cheap, clean hotel with a bathroom. They found that inns run by the People's Refreshment House Association were usually very good because they were inspected.[138] The hapless female driver was a staple of *Punch* and of magazine letter writers, and views such as those of 'E. M.' were probably typical: 'Whenever I see a lady driving ... I invariably skirt her as fully as the kind nature of the road will allow. Ladies are fascinating creatures in their proper sphere; but in the case of a car, their proper place is elsewhere except the driver's seat – purely photographic purposes excepted'.[139]

A traditional representation remained, of middle-class women driving a motor car in genteel circumstances. The flower shows played a part here, where (usually) women drove motor cars bedecked with flowers, including being attached to the wheels; this called for a slow-speed and gentle parade. Eleanor Sharp, by then Mrs Edge, was featured at the French Charity Flowers show in June 1902, and again in a Napier posing

---

[132] Smith, *Some Notes about the Anerley B.C.*, p. 49.
[133] Reference from Jungnickel, *Bikes and Bloomers*, p. 68.
[134] Brendon, *The Motoring Century*, p. 48.
[135] *The Cyclecar*, 27 November 1912, pp. 20–1.
[136] *The Motor*, 7 October 1903, pp. 200–1.
[137] *The Cyclecar*, 11 December 1912, p. 86.
[138] 'A Cyclecarist's Wife', 'Cheaper and better inns', *The Cyclecar*, 19 February 1913, p. 352.
[139] See *The Motor*, 24 February 1914, p. 141; with responses by, for example, 'E. M.', 3 March 1914, p. 192.

for her part in the Battle of Flowers at Earls Court (she won the first prize, a 'beautiful silken banner').[140] Mrs Edge was also the subject of a feature in the motoring press in her Gladiator, relating how effortless motoring was,[141] while in an advert for the same car, she explicitly linked the brand with women, drawing on a quote from *Country Life*: 'peculiarly suitable for ladies who aspire to drive'.[142] Edge also arranged for Dorothy Levitt to compete in a Gladiator for the 1903 Reliability Trials for Motor-cars, which ensured photographs and column inches in the coverage.[143]

With a few exceptions – notably Dorothy Levitt in her *The woman and her car* – an image prevailed of the inadequacy of women in the face of the technical challenge of the motor: wondering if motoring was now a lady's sport, Henry Waymouth Prance (*c.* 1882–1959) reasoned that the woman's car should be 'foolproof'. Certainly, electric cars, with their inherent simplicity and reliability, were usually associated with the woman.[144] Electricity, it said, always suggests itself as '*the* power' for ladies but electrics are generally inadequate these days. Motor-cycles 'can hardly be considered a lady's vehicle'.[145] *Motorcycling* magazine differed, where an article by the columnist Mrs C. C. Cooke, 'Motorcycling for Women', pointed out how easy it was to motor-cycle, and gave advice on how to start. Another article by the same writer, 'The Fashion of the Future: Increasing popularity of motorcycling for ladies', offered case studies of several women motorcyclists, including Mrs Thomlinson, 'a northern lady motorcyclist'.[146]

However, as the motor car and motor-cycle moved towards a standardization and with better reliability, magazine editors were recognizing the increased role that women were playing in motor traction and their potential as ambassadors for the sport. In his new column, 'Motoring for ladies', Prance decided that there was no reason why women should not be equal to men, and that one day we might even see a 'lady chauffeur'. Women, he said, now look on the car with 'less awe' and 'they deserve every encouragement'.[147] Magazines therefore occasionally published a 'woman's special', with covers bearing captions, for example, 'the lady at the wheel'.[148] Editorials such as 'The lady driver' in 1914 anticipated: 'This is to be the year of the lady driver ... she has tasted the delights of motoring as a passenger; now she is going to take control herself ... The modern motorcar, simple, easy to drive and control, clean and reliable, is now a vehicle which a lady can manage without difficulty.'

---

[140] See photo of Mrs Edge in Duncan, *The World On Wheels*, ii, p. 840; *The Times*, 17 June 1902, reported on Mrs Edge achieving first prize in the 'Paris in London' fête, her car 'decorated with Gothic arches of peonies, jonquils and marguerites'. This competition drew in competitors such as Jarrott, and contributions by *The Graphic*; *The Gentlewoman*; and *The Car Illustrated*.
[141] For example, *Motoring Illustrated*, 23 August 1902, p. 294.
[142] 'Nothing so good at the price', *Country Life*, 6 December 1902, ads.
[143] *The Motor*, 23 September 1903, coverage (and photo) starting on p. 153.
[144] See, for example, the photograph of the queen in her electric car in the grounds of Sandringham, and that of the genteel lady boarding her electric brougham: Harmsworth, *Motors and Motor Driving*, frontispiece, p. 309.
[145] H. Waymouth Prance, 'Motoring for ladies', *The Motor*, 2 May 1905, p. 350, a new column.
[146] *Motorcycling*, 14 June 1910: Article (3rd of 3): Mrs Cooke, 'Motorcycling for Women', how easy, how I got into it etc.; Mrs C. C. Cooke, 'The Fashion of the Future: Increasing popularity of motorcycling for ladies', *Motorcycling*, 18 October 1910, pp. 578–80.
[147] See also Prance, 'Motoring for ladies'.
[148] 'Ladies' special edition', cover: 'The lady at the wheel', *The Motor*, 30 June 1914.

But the precise place that women occupied as part of the motor-traction movement is profound and multilayered, if difficult to locate. Reports of women at the wheel tended to be of the exceptional, such as female racing drivers, or of record holders. Dorothy Levitt is a case in point, as the photogenic (and, no doubt, skilled) driver of high-speed Napier cars and motor-boats; Edge spotted early on the potential in promoting his businesses by arranging for photo opportunities for her. Appearing at the Automobile Club's light-car trials in 1904 as the only female competitor ensured wide coverage, and as a land-speed-record holder and racing driver, she brought no little glamour to the proceedings. She could be sure of coverage for subsequent activities, such as when she was preparing for her London-Liverpool-London trial, a 400-mile trip and the first solo female driver to do so.[149] This, and the successful conclusion of her trip, ensured favourable publicity for De Dion too (for which Edge was director). In other cases, a Miss J. Larkins was featured, with photo, having undertaken a five-day tour on her 8hp Wolseley – she thought it 'perfectly ridiculous' that a lady was unable to manage a car by herself, but ('she added facetiously') a man is useful when dirty work has to be done.[150] Similarly, Miss Alice Hilda Neville is featured as the apparently successful businesswoman who hired out, and was available to drive, her cars, and was photographed in her overalls changing tyres.[151]

A broader look at the Cheshire registrations suggest very few women indeed were registered owners of motor cars or motor-cycles to 1914, and a more likely picture would be of women using motor cars and motor-cycles registered in another's name. This is borne out by 'Mechaniste's' regular column, 'The lady on the car', in *The Motor*, which pointed out how driving had become a popular occupation for many ladies who now left their chauffeurs at home. Motoring was quickly becoming a 'necessity' for women who had to work and those who were leisured. Women who worked 'are now a vast band'. 'Lady doctors, consulting gardeners and poultry farmers' would find the car a great assistance. It continued that women do not have to know technical details, and can be confident when taking the car out without a chauffeur so that if it does break down, the network of motoring organizations and scouts will effect help. And those 'who cannot afford a man of any sort to look after the car, or merely a boy for washing or cleaning' need not be deterred but should get a few lessons in maintenance from a motor engineer.[152]

Conclusions

Jarrott did not forget his ride in Number 5 in 1897. It wasn't as if he had enough to contend with by nursing an utterly unreliable motor vehicle; his passenger Wellington had brought a revolver along and took to shooting at rabbits as they drove, and this

---

[149] *The Motor*, 4 April 1905, p. 246.
[150] *The Motor*, 28 February 1905, p. 108.
[151] *The Motor*, 20 May 1913, pp. 735–6; 2 September 1913, p. 193, with a picture of her 'in the garaging and hire business. She is a capable driver, and is evidently a competent business woman'.
[152] *The Motor*, 11 March 1913, pp. 252–3.

excited the attention of a mounted police patrol which they had to outrun. Compare this experience with Jarrott's musings in his book of 1906: 'the romance of motoring exists no longer', he said. One can now matter-of-factly drive from London to Monte Carlo in forty hours, 'practically certain' of arriving within a given time, and a non-stop run from London to Edinburgh is within the capabilities of 'most well-built cars'.[153] Without this wholesale shift in reliability, motor traction would have remained a hobbyist curiosity for historians to puzzle over today. The entrepreneur (Jarrott included) played a part here in bringing an increasingly viable consumer durable to the public, but the role of the consumer in shaping this product should also be recognized. In that space of nine years described by Jarrott, the way the motor car was perceived and consumed changed fundamentally.

The entrepreneurs in those early days knew, of course, that their products were challenging to sell, with no brands, no reliability, no infrastructure, and little demand. Edge, in his later years, described motor cars prior to 1900 as 'absolute porcupines, all prickles!'[154] But one advantage he, Jarrott and others had, no matter how little they knew about what they were selling, was that they knew more than the customer; this was apparent when Jarrott was explaining motor traction to the public at the 1896 Stanley Exhibition.[155] The product, then, was 'immature', leaving customers, for example, fretting over the fundamentals, such as where the engine should be positioned on a motor-cycle. Indeed, should they be buying something, such as a motor-cycle, which might turn out to be just a passing fad? The same customers would worry about what kind of motor car they might buy: should it be petrol, should it have pneumatic tyres? Should the passengers face backwards or sideways (why not? they did in a horse and cart)? Magazines, keen to keep readers engaged, were quick to play up improvements in motor traction, to comment on how 'dreadful' the motor car had been just a few years earlier. Magazine writers 'identified' a clear trajectory: the cyclist naturally becoming the motor-cyclist, and then in turn the motor-car driver. It was useful for the writers to create the idea of a fraternity, of sensible people making rational purchasing decisions, all cooperating in mutual support. The consumer in turn fed in to the process by making suggestions about what was lacking in products then available, by building or adapting their own and sharing these ideas. They specified what they wanted and how much they would pay for it. They grumbled about the 'customer experience' when dealing with agents.

And by turns, by about 1905, the purchaser was offered products with a degree of standardization – most motor cars now had the engine at the front, and that engine was probably a petrol engine. Seats were facing forward, while wheel sizes, front to rear, were now the same. Alternatives remained, such as the means of propulsion (steam, electric); the method of transmitting the drive (belt drive, chain drive, shaft drive for the motor-cyclist, and chain or shaft drive for the motorist),[156] but otherwise this was a rapidly maturing market. Choice was improving year on year, and reliability

---

[153] Jarrott, *Ten Years of Motors and Motor Racing*, p. 266.
[154] Edge, 'Looking back', pp. 7–9.
[155] Jarrott, *Ten Years of Motors and Motor Racing*, p. 8.
[156] See the cover, *Motorcycling*, 24 January 1910.

improved; the 'medical man' J. R. Ratliffe, who had wanted to be able to start his car from the seat, would have found that feature available in the cheapest of light cars for sale just a year later.[157] The 1900 Motor Show had ninety-four exhibits, while by 1904 there were 330.[158] Motoring magazines published full lists of vehicles available and their costs. The consumer was learning that further comfort and convenience came with paying extra for windscreens and hoods, and then investing in after-market accessories such as speedometers or 'Stepney' wheels. At the same time, guest speakers at the Automobile Club were reassuring their audiences that motoring was dreadfully expensive, while *The Motor*'s message was quite different, seeking to suggest that motoring was affordable.

The appearance of 'old timer' networks was also symbolic of the change in consumer patterns. The 'old brigade', the first generation of motor-traction entrepreneurs, needed to show they could bring experience, wisdom and glamour from the old days, to make themselves relevant in a market where they had once sold what had become 'old crocks'. By the cyclecar boom of 1910 (described in Chapter 7), there were many new entrepreneurial names joining the fray and the first generation was starting to move on or be edged out. Edge was associated with the old and the new; he was in 'old timer' societies, he served as president for a variety of cycling, motor-cycling and motoring clubs, he endowed prizes and awards for these clubs just as his mentors such as R. H. Fry had when Edge was a young club member. Edge continued to have influence beyond the First World War in the cycling and motoring worlds but by then his Midas touch was starting to elude him.

In this man's world, women, as potential consumers, were recognized as an avenue of revenue. Due to the use of pseudonyms by letter writers, it is difficult to get a handle on the place that women played in the motor-traction movement, and there is a tendency to be 'blinded' by the glamorous and successful motorists such as Dorothy Levitt. The social mores of the period meant a wider and inevitable resistance to women who motored, had motoring businesses, who raced, just as those who had cycled, or wore rational dress when doing so. But equally, the (now) patronizing language ('so easy even a woman can drive it' and so on) also served as a universal metaphor, understood to mean it was easy for men too. In an age when thumbs and arms were so easily broken by a recalcitrant starting handle, this counted for much.

The old brigade, steady and careful artisans, women – whoever these consumers of the new motor traction were, and however they had come by that world – had been ensnared by the promise of the 'open road', and it is the pursuit of this, and the creation of different types of 'tourists' in the process, that we turn to next.

---

[157] See the Crestmobile advert, *The Motor*, 8 July 1903, ads. Such mechanical solutions were suited only to small engines and remained comparatively rare until the electric starter appeared.
[158] *The Motor*, 15 March 1904, p. 163.

6

Tourists

'E. M. B. P.' very much enjoyed his motoring tour of the French and Italian Alps in 1910. He and four others spent three weeks driving their Siddeley 14/20hp from Dieppe in France to Aosta in Italy, and back, covering 1,600 miles. 'That this was an inexpensive one [holiday] all will agree, [he wrote, calculating £15 16s per person] and it served further to impress on us all that a motoring holiday on the Continent is worth two spent in England, where the police traps on the open road and the expensive hotels in the towns render the motorist's lot by no means a happy one.' He had taken the trouble to bring photographic equipment to record the event, and *The Autocar* was happy to publish his story.[1]

It has been shown how, through the energies of the motor-traction entrepreneur, a 'consumer' emerged who increasingly situated himself as quite distinct from the first pioneering motorists. This consumer expected reliability, to set off on a journey and arrive without any fuss, whereas for the pioneer, the struggle of the journey was all. This chapter will suggest that a significant part of the attraction for the motorist (and the cyclist before him) was the tour, the being on the road. While the dream of the 'open road' was sold by the entrepreneurs as a reason for buying into motor traction, the tourists in turn fed back through letters to the motoring press their comments, criticisms and information. That, with the appearance of books on cycle- and motor-touring, helped further shape the experience and expectation of the tour for others.

Private touring had been widely undertaken, first by cyclists in the 1870s and then by private motorists since the mid-1890s. This is known now through accounts published by the Cyclists' Touring Club's *Monthly Gazette* and the wider cycling press from the 1880s, and then the motoring press, starting with *The Autocar* in 1895. These tales had an appeal beyond their own hobbyist press; Henry Sturmey judged that there was sufficient public interest, for example, to publish as a collected edition the notes on his tour from John O'Groat's to Land's End in 1897.[2] The Automobile Club conducted its first successive-day tour for members over Easter 1898, an event requiring much preparation and a wisely suggested forty-mile limit per day,[3] and the local motor clubs were soon doing the same. The first cycling and motoring clubs initially undertook their tours on a Saturday, so that they could be seen to keep Sunday sacred. These tourists

---

[1] *The Autocar*, 15 January 1910, pp. 80–2, 108–11, 164–7.
[2] Sturmey, *On An Autocar*.
[3] Described in Nixon, *Romance Amongst Cars*, pp. 99–107.

sometimes took photographic equipment with them, or made drawings, to record their trips. This chapter will consider two types of tour – the pioneers' tour and the 'consumer' tour; then cheaper opportunities such as camping, and the consequences of touring, notably the incursion of the working classes, despoliation of the countryside, traffic and the speed trap.

Historians have considered how speed and wholly self-directed travel were becoming part of the touring experience.[4] Travel by the train in particular was 'deeply alienating' compared with the 'nobility' of the horse carriage, whose travellers could 'chart their own path'. Bierbaum had written in 1902 that 'whoever goes travelling in a railway coach forgoes, for a time, his freedom. Every trip made by rail is a transport of prisoners'.[5] Denning has described the new transportation technologies (bicycle, automobile, aeroplane) – the 'transports of speed' – and their 'transformative power'. By the First World War, bicycles, motor-cars and aeroplanes had appeared to combine 'the independence and privacy of horse travel with the power of modern machinery, allowing moderns to harness speed and conquer natural limits through human ingenuity'. Speed became 'pleasurable, thrilling, embodying power, independence and progress'. These new transports offered what Duffy called 'speed for speed's sake', allowing 'individuals to transcend modern ennui through their mastery of velocity'.[6] They were 'escapist tools for leisure and adventure for the vast majority of their enthusiasts during the fin de siècle'.[7]

The cultural historian David Jeremiah has pointed out that *The Autocar*, the first British motoring magazine, had a 'Tours and Runs' section from the outset (1895). It anticipated

> freer intercourse, unlimited action and constant change of scene, and with vehicles able to be driven ten or fifteen miles an hour for very long periods with only occasional halts for replenishing the water and petroleum spirit, there is nothing to prevent the populations of large towns dispersing into hitherto, almost inaccessible parts of the country, and enjoying the health and freedom which are often denied them in the busy haunts of men. This easy transport promises to help in solving one of the most serious problems which has been vexing the soul of the political economist for years past – the desertion of people from the country for the towns.[8]

Flicking through any motoring magazine of the time would reveal similar stories, idealized drawings, and photographs of travelling accounts of motorists on tour. Cycling magazines were the same. The 'culture of travel' had become a part of the

---

[4] See, for example, Stephen Kern, *The Culture of Time and Space, 1880–1914* (London: Weidenfeld and Nicolson, 1983).
[5] From his *Eine empfindsame Reise im Automobil*: quoted in Wolfgang Sachs (trans. Don Reneau), *For Love of the Automobile: Looking Back into the History of Our Desires* (Berkeley and Los Angeles, CA: University of California Press, 1992), p. 7, from Denning, 'Transports of speed', pp. 380–1.
[6] Duffy, *The Speed Handbook*, cited in Denning, 'Transports of speed', p. 381.
[7] Denning, 'Transports of speed', pp. 380–91.
[8] Jeremiah, *Representations of British Motoring*, p. 14, from *The Autocar*, 28 December 1895, p. 107.

popular consciousness by the end of the nineteenth century.[9] Jeremiah has shown how motoring, since the 'success' of the Locomotives on Highways Act of 1896, 'became increasingly identified with the British countryside'. He continued:

> Central to the construction of the pleasures of motoring was an imagined relationship that established the idea that the motorcar was reliable, able to deal with every eventuality; offered freedom, with the road to yourself, and the ability to discover the hidden beautiful Britain.

This was a 'reopening' of the British countryside where 'rural motoring was packaged as a consumable'.[10] Increasingly so between the Wars, motorists were encouraged to see their own country first, while the motoring press, just as the bicycle press had, resorted to artist-illustrators, with motor manufacturers and petrol companies using 'innovative' advertising campaigns, all to promote 'the mutual benefits of the harmony between modern motoring and the countryside'.[11] Motoring-magazine editors in the period to 1914, however, also drew on an existing culture of finding and exploring the 'open road' and urged new and intending motorists to do the same. Indeed, Coulbert has shown how the 'open road' was 'a buzz-phrase in travelogues and other motoring literature well into the 1930s. It suggested opportunity, an enticing prospect, chance possibilities, and ultimately adventure (or misadventure)'.[12] A new genre of novel, based on the motor tour, had proved popular; a review of the Williamsons' *The lightning conductor* (1903) declared it 'An ideal romance of the car' which 'expresses admirably the fascination of the open road'. *The Car Illustrated* thought it could 'be read with pleasure by motorist and non-motorist alike'.[13] Books giving guidance about travel, or how to do so on the cheap, and increasingly detailed maps, were very popular.[14] Writers such as 'Owen John' (Owen John Llewellyn [1870–1943]), C. L. Freeston (1865–1942) and J. J. Hissey wrote several motoring travel books, which Matless has recently categorized as 'motoring pastoral genre', to include guidebooks and travel books for

---

[9] J. Steward, '"How and where to go": The role of travel journalism in Britain and the evolution of foreign tourism, 1840–1914', in J. K. Walton (ed.), *Histories of Tourism: Representation, Identity and Conflict* (Clevedon: Channel View, 2005), pp. 39–54, at p. 39.

[10] David Jeremiah, 'Motoring and the British countryside', *Rural History*, 21:2 (2010), pp. 233–50, at p. 233.

[11] Jeremiah, 'Motoring and the British countryside', p. 234.

[12] Coulbert, '"The romance of the road"', p. 208.

[13] *The Car Illustrated*, 3 December 1902, p. 44.

[14] For example, J. E. Vincent, *Through East Anglia in a Motor Car* (London: Methuen, 1907); Hissey, *Untravelled England*; H. Massac Buist, *Motoring to Stonehenge* (London: 1907); John Dillon, *Motor Days in England: A Record of a Journey through Picturesque Southern England with Historical and Literary Observations by the Way* (New York: G. P. Putnam's Sons, 1908); Owen Llewellyn, *The South-Bound Car* (London: Methuen, 1907); H. V. Morton, *In Search of England* (London: Methuen, 1927). See also, Alun Howkins, *The Death of Rural England: A Social History of the Countryside* (London: Routledge, 2003). For maps see T. R. Nicholson, *Wheels on the Road: Maps of Britain for the Cyclist and Motorist, 1870–1940* (Norwich: Geo Books, 1983). A useful example of a period guide book is Charles Howard, *The Roads of England and Wales: An Itinerary for Cyclists, Tourists and Travellers* (London: George Gill and Sons, 1882), which was in its fifth edition by 1897.

Figure 6.1 Motor cars at Ascot, 1900, the first year they were allowed into the enclosure. Hedges Butler's motor car is to the left. The photograph was taken by Argent Archer. From Frank Hedges Butler, *Fifty years of travel by land, water & air* (London: T. Fisher Unwin Ltd., 1920), p. 115.

motor travel around England.[15] The roads they used, and the politics of access to them, is also engaging the attention of historians.[16]

Furthermore, Buzard has drawn attention to the anglicization of continental destinations during the nineteenth century, where English food and newspapers could be found, while magazines bemoaned and caricatured the English tourist abroad, following the same circuits and using the same guide books.[17] Coulbert has spotted the common theme for most travel writers of the time: a preoccupation with industrialism, idealizing pastoral landscapes, and class prejudice against working-class tourism.[18] The desire to travel to beauty spots, as a couple or as a group, was being stoked up by magazine editors who could see how travel accounts made for higher circulation. The best illustrators, notably Frank Patterson (1871–1952), depicted and promoted the destination of the unspoilt countryside scene, free from the noise and smoke of the city.[19] And for those who could afford it, such as 'E. M. B. P.', an ever-larger body of literature was available to help inform about the tour abroad.

[15] David Matless, *Landscape and Englishness* (London: Reaktion Books, 1998), p. 63, cited in Martin Walter, '"The song they sing is the song of the road": Motoring and the semantics of space in early-twentieth-century British travel writing', *Transfers*, 5:2 (2015), pp. 23–41, at p. 23. For Freeston, See Richard Fletcher, 'Charles Lincoln Freeston (1865–1942): Transport writer and journalist', *Aspects of Motoring History*, 14 (2018), pp. 29–40.

[16] For example, Keith Laybourn with Taylor, *The Battle for the Roads of Britain*; Joe Moran, *On Roads: A Hidden History* (London: Profile, 2009); Reid, *Roads Were Not Built for Cars*; Plowden, *The Motor Car and Politics in Britain*.

[17] James Buzard, *The Beaten Track: European Tourism, Literature and the Ways to Culture, 1800–1918* (Oxford: Clarendon Press, 1993), chap. 2, esp. pp. 89–93.

[18] Coulbert, '"The romance of the road"', p. 201.

[19] *The Motor*, *passim*, but also see Jeremiah, *Representations of British Motoring*, p. 21: the Frank Patterson drawing of 'Harting and the South Downs', 1906.

The travel writer J. E. Vincent (1857–1909) wrote about motor touring. He decided the only limits a motorist could experience came down to personal choice and physical endurance, 'and he need never be troubled, as the horseman must be, from time to time, by doubts whether his pleasure may not be causing pain to the organism which carries him willingly from place to place'.[20] (This was also a sales method, to emphasize that a motor vehicle could travel all day, whereas it would be cruel to make a horse do this.) Hissey claimed never to follow a planned route and delighted in his chance discovery of rural locations.[21] The journalist Filson Young would have agreed. 'The mere driving of a motor-car, however beautiful the country may be', he wrote, 'is a small and decreasing pleasure if the beginning and the end of the drive are one and the same point'.[22]

'Real' tourists

There was, first of all though, what the pioneers – Jarrott's 'old brigade' – might have called 'real' touring. For these pioneering tourists, appalling roads and public hostility were normal; breakdowns and unscheduled stops were all part of the journey. They were travelling much further and faster than any road-user before, and finding the roads and infrastructure unfit for their new purposes: signage, junction design, road camber, road 'discipline'. These had never been a 'problem' before, but were all becoming bones of contention. These tours were initially therefore conducted at an unpredictable pace; it was the tour itself that was the adventure, not necessarily the arriving. The spirit of the middle-class amateur sportsman prevailed: it was, then, the taking part – the journey – that mattered.[23] A flurry of autobiographies by these first sportsmen describe an immersion in a culture of adventure, of overcoming obstacles. The sense of character formation and stoic heroism was portrayed by fictional characters created by authors such as H. Rider Haggard (1856–1925), and then made real by the adventuring and exploring of the likes of Frank Hedges Butler (1855–1928) and Montague Grahame-White, all adopting the bicycle, motor traction, or aviation.[24] G. H. Smith said of the first cycling-club runs, 'on those old runs they really were exploring', and many dozens of articles published in the earliest motoring magazines featured the stubborn determination of the pioneer motorists to overcome the difficulties presented by machine, terrain and unfriendly natives. Often, the pioneer tourist made sure a camera was on board, so that the adventurer could be captured on

[20] Coulbert, "'The romance of the road'", p. 205.
[21] Coulbert, "'The romance of the road'", p. 205.
[22] Filson Young, *The Happy Motorist*, p. 26.
[23] This spirit is apparent in the pioneers' reminiscences; a useful collection is in [Lord] Montagu, *The First Ten Years of Automobilism, 1896–1906*.
[24] I have in mind the hero of H. Rider Haggard's *King Solomon's Mines* (1885), Allan Quartermain. For background on Victorian imperialism see Steinbach, *Understanding the Victorians*, pp. 65–72, esp. pp. 65–6, 71. Frank Hedges Butler was a big-game hunter, balloonist and motorist: see Frank Hedges Butler, *Fifty Years of Travel by Land, Water, and Air* (London: T. Fisher Unwin Ltd, 1920); while Montague Grahame-White was a racing driver, playboy and potential spy: Grahame-White, *At the Wheel Ashore and Afloat*.

film facing and overcoming the hardships. Additionally, professional photographers were engaged to cover events, most notably Argent Archer for Automobile Club tours such as the Thousand Mile Trial of 1900.

The hardships faced by the earliest cycle tourists have been recorded in cycle-club histories and diaries. The first Anfield Bicycle Club rides, in about 1879, required 'tough' riders, who 'at any moment might be thrown over the handles by a stone (of which there were plenty), or by riding downhill', and tyres were of solid rubber more to protect the wheel rim than the rider 'from the unspeakable roads'. The rider, then, 'grew proud of his aching bones'. The club was joined by Lawrence Fletcher (b. c. 1861) from the CTC, whose idea of a Saturday ride in 1879 was 227 miles in twenty-four hours from Liverpool and covering much of north and mid Wales.[25] G. H. Smith's diary documented his own cycling tours in the 1890s with the Anerley Bicycle Club and gives an insight into the hardships that the first cycle tourists would expect. Smith, like many of his peers, recorded each trip assiduously, tallying the mileage for each year.[26] Writing the club history in the 1930s, he described his experiences as a club cyclist in the 1880s and 1890s in more detail. There was a club run, every Saturday, meeting at the Robin Hood pub, which was 'not at first very lengthy', only roaming once a month so far as to have tea out. But most rides were events: he reported having had a 'violent cropper', leaving him 'much shaken',[27] while S. F. Edge and his new (first) wife Eleanor had a 'cropper' on their tandem.[28] *Bicycling News* recalled in 1900 how '[i]t was the early prejudice which caused the formation of cycling clubs in various localities, for it was almost necessary to ride in company in order to insure fair play from horsey people'.[29] Edge reported a man who 'tried to run us down',[30] while a 'boy upset Fraser [so] we run him in'.[31] The reality of cycling on a summer bank holiday night, 'dark' and 'roads full of drunks' can only be imagined.[32]

Existing signage was 'inconveniently high for [motor-car] drivers', decayed or half-hidden, with 'small lettering scarcely legible at speed'. Signs tended to identify the nearest village, rather than towns further afield.[33] The roads in the southern counties were sandy, gravel or chalk. They were dusty if the wind blew, and if the dry weather was prolonged, 'the sand roads became terribly loose and cut up', causing the Brighton Road to be impassable for cyclists beyond Purley.[34] This, with constant tyre trouble, meant the early cyclist not infrequently had to get the train home. In the winters, mud

---

[25] [anon.], *The Black Anfielders*, p. 3.
[26] 1891: 3,992 miles on 115 rides, 'a very wet and windy year, the roughest riding year experienced'; 1892: 4,660 miles on 131 rides; 1893: 4,461 miles on 126 rides, 'a magnificent riding year'; 1894: 4,760 miles on 118 rides; 1895: 5,066 miles on 117 rides. All from Diary of G. H. Smith.
[27] Diary of G. H. Smith, 26 July 1891.
[28] Diary of G. H. Smith, 1 July 1894.
[29] *Bicycling News and Motor Car Chronicle*, 20 June 1900, p. 16.
[30] Diary of G. H. Smith, 28 February 1891.
[31] Diary of G. H. Smith, 13 March 1892.
[32] Diary of G. H. Smith, 6 August 1894.
[33] Kathryn A. Morrison and John Minnis, *Carscapes: The Motor Car, Architecture and Landscape in England* (New Haven, CT: Yale University Press, 2012), p. 239.
[34] G. H. Smith, *Some Notes about the Anerley B.C.* (privately pub., 1930), pp. 6, 9. Roads in the metropolis are described by Turvey, including the dust when dry, the dirt ('slop') when wet, the noise, the different road construction techniques tested, with their costs and longevity weighed up, and the absence of any parking problems: R. Turvey, 'Street mud, dust and noise', *London Journal*, 21:2 (1996), pp. 131–48.

ruts would be turned by frost into 'iron rails', which meant the road had to be walked. In the entry for 10 January 1891, he wrote: '[cycled to] Riddlesdown on my solid[-tyred "safety"]. Very cold, hard frozen snow', on which he did eighteen miles (the following day, the Thames froze over). Descriptions of the road varied, including 'fair', 'good', 'heavy', 'bad', 'grand', 'hard and rough', 'awful', 'utterly rotten' and 'terrible stones' (which cut the tyres). The road from Arundel to Offington on the club ride in July 1891 was so bad that Edge's father Alexander gave up. As a consequence of these conditions, many club runs were short, and most races only ten miles or so. Writing forty years hence, 'To ride home from Riddlesdown on a dark winter's night, with ruts, mire, heaps of metal for machines [bicycles], and strips of rubber for tyres, was quite far enough for most of us'. On another occasion, one of the club came 'on horseback' but fared no better than the cyclists, the horse finding the ice difficult.[35] 'Horsey people', said Smith, slashed at him with their whips, while small boys hurled stones or tried to thrust sticks between their spokes.[36] Dogs, even fox hounds, were liable to hunt the cyclist: 'The writer once had a whole pack in full cry after him and only escaped a nasty mauling by dismounting and standing quite still until the huntsmen arrived and flogged the hounds off'.[37]

Wayside hotels and inns, now 'so smart and prosperous' were then, 'mostly, in the last stages of decay'. Despite a want of custom, publicans 'too frequently gave no welcome to the cyclist but treated him as a foreign and probably dangerous person; and often have we been refused food and shelter on the patently false plea that they were full up'. This attitude helped create a camaraderie amongst the peer groups, with fond and long-held memories of those inns willing to welcome the cyclists; these include Mrs Dibble on the Ripley Road, and the Cricketers pub where Alfred Nixon stood the riders a bowl of punch.[38] While the fare in the south was 'generally indifferent', in the north it was 'generous and very cheap'; the first 'north' tour at Easter 1889 had reasonable 'fare and bills' such that the tourists could live for a 'very few shillings a day', but the weather and roads were so bad they did not do another.[39]

While the club offered mutual protection and support, it also created a brotherhood, united by their determination to tour as cyclists. Smith described how cyclists, on meeting others, would stop and chat and exchange information.[40] Cycling and motoring articles increasingly referred to a 'fraternity', of like-minded souls. In the event of a breakdown or other morale-sapping event, heart-warming accounts encouraged them on, to let them know they were not alone; a cover illustration of *Motorcycling* in 1910 showed one motorcyclist stopping alongside another, stricken at the roadside, asking 'Can I help you?'[41] In one story of motor-cyclists enduring indifferent reliability in 1910, the conclusion was '[we] do not by any means think

---

[35] Smith, *Some Notes about the Anerley B.C.*, pp. 13, 14; Diary of G. H. Smith, 10 January 1891; Diary of G. H. Smith, 5 July 1891.
[36] Smith, *Some Notes about the Anerley B.C.*, pp. 6–7.
[37] This was when cycling from London to Manchester on solid tyres: 'Pillars of the pastime no. 2: G. H. Smith', *Cycling*, 1 November 1917, p. 312.
[38] Smith, *Some Notes about the Anerley B.C.*, p. 15.
[39] Smith, *Some Notes about the Anerley B.C.*, p. 21.
[40] Smith, *Some Notes about the Anerley B.C.*, pp. 6–7.
[41] *Motorcycling*, 5 July 1910, cover. The caption continued: 'The bon camaraderie amongst motorcyclists is one of the most notable features of the pastime'.

of "giving up motorcycling".[42] Smith recounted the popularity of the club runs in the late 1880s: fifty went out on a run in October 1889, while eighty-five sat down for tea at the Rose and Crown in Riddlesdown, Surrey, in October 1891, and one hundred turned up for the ride to the same village in February 1893. There was a tendency to ride 'in a mob', with 'young fellows swinging along in one mass'.[43] As the tourist went further afield, the fraternity played an increasingly important part in facilitating the tour. It has been shown how clubs had influential patrons, such as MPs, JPs or baronets, which meant where an introduction was required, or the intervention of a person of influence, the mention of the club would open doors. Smith pointed out in his history of the ABC how the very name of the club carried weight when travelling abroad. Going to France in 1894, for example, in order to ensure smooth passage of bicycles at Dieppe, the British Consul advised that if they all wore the ABC badge, 'all would be well', with no difficulty or delay at customs.[44]

But there was also the informality of the private tour indulged by friends, or alone – say, to Scotland for a week, or simply cycling in the local area. Smith's diary gives a window into this kind of touring; he frequently started a run with the club and then splintered off. He described his solo tour of Scotland in May 1891, when he spent a week in the Highlands ('very wild country', but the roads were 'fair', if rough in parts).[45] The tours were intended to pack as much in as possible. Cyclists from the outset went on rides, often overnight, staying at inns.[46] To maximize time on the tour, it was normal to ride in the dark, or overnight; this was possible without street-lighting because the unsealed roads tended to be white or light grey in colour, thus reflecting the moonlight.[47] It was a regular practice, because Smith commented on, for example, the 'grand night', or the 'fine moonlight'. A week's tour in May 1894 finished with him getting home at 1.15 am[48] (and presumably back to work that same day). Very early starts were also part of the experience: Smith described a 4.30am start for a North Road twenty-hour race.[49] It was newsworthy, but soon nothing special, for cyclists to complete the Land's End to John O'Groat's. One such, Dr Horace Mansell, practised for his tour in 1888 by cycling around Islington, London; going from London to Land's End, then to John O'Groat's took sixteen days, and he calculated his expenses at £11 13s 10d.[50] Continental and even world tours were also fairly common for those with the time and money; Smith left off writing his diary in 1896 to complete a world tour by bicycle.

A swapping of machinery was also usual. Smith borrowed Edge's 'pneumatic', and on another occasion his tandem.[51] He had access to 'safeties', tandems, and tricycles,

---

[42] 'A night of misadventures', *Motorcycling*, 3 January 1910, p. 190.
[43] Smith, *Some Notes about the Anerley B.C.*, pp. 12–13: refs to 26 October 1889 run; Diary of G. H. Smith, 31 October 1891; 25 February 1893 (Riddlesdown).
[44] Smith, *Some Notes about the Anerley B.C.*, p. 25.
[45] Diary of G. H. Smith, 6 May 1891.
[46] Smith, *Some Notes about the Anerley B.C.*, p. 20.
[47] Nick Clayton, 'The birth of Tarmacadam', *Cycle History*, 24 (2013), pp. 88–92, at p. 91.
[48] Diary of G. H. Smith, 14 May 1894.
[49] Diary of G. H. Smith, 10 September 1892.
[50] 'An imperfect account of a "Safety" cycling tour taken during my summer's holiday', August–September 1888, unpub. travelogue in 5 vols, auctioned 26 March 2019 by Lawrences, lot 398.
[51] Diary of G. H. Smith, 28 June 1891, 4 October 1891.

variously with pneumatic tyres or 'solids'. Edge, as a 'maker's amateur' and salesman, would have had access to the latest equipment. The appearance of the 'safety' from the late 1880s did not necessarily cause a wholesale adoption of the new machinery, and some of the early cyclists persisted with their 'ordinaries', Faciles or Kangaroos;[52] see for example, Montague Napier choosing to ride his 'ordinary' into the 1900s, and a photo of three male touring cyclists, all on venerable bicycles, taken in about 1894 when the 'safety' was much the most popular style.[53]

The rewards of early touring, though, were priceless, and, as Smith realized, fleeting: 'Perfect quiet' out in the country, 'miles would be covered without meeting a vehicle'. 'These early runs [. . .] really were exploring', he wrote. 'Every bend in the road was pleasant speculation' [and] 'it was the uncertainty as to the state of the roads and the certainty of some of the riders having headers that made each run an adventure and a great novelty'.[54]

Charles Jarrott, for one, came from just such a cycling background,[55] and extended his experience into motoring from about 1896; his first motor-car trip from Margate to London in 1897 with Frank Wellington has been described in Chapter 5. This was an experience which might have put many off motoring, but simply hardened his resolve to motor more.[56] So when Jarrott wrote of the 'golden age' of motoring, an age ended by 1906, he was in earnest. He bemoaned the demise of the old touring:

> I have so often heard it said by motorists who have followed the sport and pastime from its commencement that motoring to-day presents few charms, gives little pleasure, and small amusement, because, owing to increased speed and reliability to which motor-cars have attained, incidents of an unexpected character never occur, and a journey, either long or short, resolves itself into a mere passage from one place to another.[57]

Mom's 'adventure machine',[58] then, was continuing to evolve into a reliable light car. Motoring was ceasing to be a sport. Jarrott reckoned that for many, driving had become monotonous. At one time, his friends (the fellow pioneers) had been prepared to undertake 'all little duties necessary for a run', the work 'a labour of love', but now have given up driving and are driven. The car had become 'a means of conveyance' instead of a sport despite being more reliable and much easier to drive.[59] St John Nixon

---

[52] Smethurst is very useful in trying to explain and theorize the development from 'ordinary' to 'safety', taking in 'short-lived transitional machines' such as the Kangaroo or Facile: Smethurst, *The Bicycle*, chap. 1, esp. pp. 41–9.
[53] A posed photograph of three male cycle tourists in *c.* 1894 shows only one using a 'safety': *The Boneshaker*, 206 (2018), p. 64 (the detail and location of the photograph were discussed by readers over successive volumes).
[54] Smith, *Some Notes about the Anerley B.C.*, pp. 6–7, 9–10.
[55] He mentions his love for cycling and cricket prior to 1896: Jarrott, *Ten Years of Motors and Motor Racing*, p. 6.
[56] Jarrott, *Ten Years of Motors and Motor Racing*, pp. 13–18.
[57] Jarrott, *Ten Years of Motors and Motor Racing*, p. 25.
[58] Mom, *Atlantic Automobilism*, chap. 2.
[59] Jarrott, *Ten Years of Motors and Motor Racing*, p. 277.

recalled dropping in on fellow pioneer Mark Mayhew (1871–1944) in the 1930s for 'having a chat about motoring when motoring was a sport and not a road menace'.[60]

So, as with cycling, hardship and discomfort were integral components of the pioneering motoring experience. And so often, they found themselves acting as ambassadors for the 'movement', determined to arrive – and be seen to do so – lest the onlooking masses remained unpersuaded of the movement's merits and reliability. The basic infrastructure was not in place. The difficulties of J. A. Koosen in obtaining petrol have been discussed; this was clearly a significant hurdle as so many pioneers mentioned it. For Henry Hewetson's first car, an £80 Benz he imported from Germany, he too was unable to obtain petrol and had to use benzoline from oil shops. He recalled that the specific gravity was so heavy that he had to heat it in cold weather.[61] Accounts relating to the Thousand Mile Trial of 1900 underline the difficulties (in-depth arrangements well in advance had meant petrol was available at all the stops. But even then, in Manchester, petrol could be bought in quantity only in two places). On the day the Trial was leaving from Leeds, a 'hurricane had been blowing', yet shortly after 5am the motorists left their warm hotels 'prepared to undergo fatigue and sacrifice themselves in storm and rain in order to convince the British public of the delights of automobilism. Some may have wavered; a few quailed before the elements; but none confessed to such weakness'.[62]

It took the passage of thirty or forty years before Smith could be candid about the earliest days of cycling. One example is his description of a couple riding a 'sociable' [a tricycle, two abreast]: the sociable was anything but – 'anything more liable to generate unsociable feelings than two hot and tired cyclists trying to get a little pace out of these crushing masses of heavy tubing and wheels, would be hard to find'.[63] But in the process of taking 'ownership' of the road, those cyclists were pioneers in rebuilding what was a moribund network out of towns, in challenging hotels, inns and halfway houses to improve standards. This process was accelerated by the earliest motor-car user, requiring accommodation (garaging) for the motor car and for the entourage.

The 'consumer's' tour

The coming of tolerable reliability, and an increase in the number of second-hand vehicles available, marked the turning point for the evolution of the 'consumer's' tour. Increasingly, the motor-tourist cared less for the mechanicals, beyond what maintenance had to be done. Filson Young had identified them: 'The happy motorist is he who has the freedom of the roads without being in bondage to a pastime or a fad', that is, the new motor tourist should be above getting absorbed in the mechanical details.[64] Up-to-date maps were now important, particularly those with contours, useful for informing

---

[60] St John Nixon, 'Myra Edge – Appreciations', *VCC Gazette*, 8 (1969), p. 432.
[61] Henry Hewetson in Montagu, *The First Ten Years of Automobilism*, p. 54.
[62] *Motor Car Journal*, 12 May 1900, p. 172.
[63] Smith, *Some Notes about the Anerley B.C.*, p. 5.
[64] Filson Young, *The Happy Motorist*, p. 11.

the tourist of potentially difficult routes and the likely refreshments and spectacles on the way; these had not been at the disposal of the earliest cycle- and motor-tourists, although books describing road conditions and recommending hospitality had started to appear in the 1880s.[65] Some travellers put up with what they could get: J. E. Vincent, for example, observed that the Murray guides he used on his motor tour of East Anglia in 1907 were thirty years out of date.[66] A series of maps advertised in 1907 showed their use for 'touring: reliable and inexpensive', and typically cost one shilling (or about two shillings if on cloth); they covered a range of scales, from twenty to 150 miles around London. In addition, Bartholomew maps were advertised as based on the Ordnance Survey, at a scale of half an inch to the mile, with thirty-seven sheets covering England and Wales. Selling for one shilling (cloth, two shillings), they included main roads, secondary roads, footpaths, were coloured for altitude, had 'most useful features for cyclists and tourists', and showed inns, hotels, fishing streams, lochs, woods and forests.[67] Magazines published speed-trap maps to 'help' the motor-tourist, as discussed later. A new genre of motoring touring guide books had appeared by about 1905, by, for example, Filson Young, Max Pemberton (1863–1950) and Major Matson,[68] usually providing a 'complete' reference to how to buy, maintain and drive your car as well as where to take it on tour and how to behave once there (usually, to always show respect for the locals). Feature writers in the motoring magazines such as 'Owen John' or C. L. Freeston described their motor tours, usually in a form that emphasized ease and simplicity. Travel by motor car formed the purpose for the journey itself.

Filson Young was well qualified to write for this new tourist. A well-connected journalist and author, he was known to *Daily Mail*-owner Alfred Harmsworth (editing its 'literary' page from 1903), and to Edge and Jarrott. In 1906 he persuaded Jarrott to drive him to the south of France, and in the 1920s he published *Cornwall and a light car* (1926), using an AC motor car supplied by Edge,[69] a company in which Edge then had a controlling interest. An adviser to the BBC in the 1920s, he interviewed Edge on BBC radio for a documentary on the 'recent renaissance of the road'.[70] Filson Young's *The complete motorist* (1904) and *The happy motorist* (1906) captured the spirit of the 'open road', very much from the point of view of the leisured consumer who 'should have no mechanical preoccupations, or his true enjoyment will be menaced'.[71] This kind of literature was much in demand as suggested by the number of titles, and subsequent reprints and editions, appearing at that time. Just as the pioneers were increasingly seen as being from a past generation, and any vehicle older than a few years condemned

---

[65] Morrison and Minnis, *Carscapes*, chap. 9, esp. pp. 239–43.
[66] Vincent, *Through East Anglia in a Motor Car*, cited in Coulbert, 'Perspectives on the road', p. 37.
[67] *Motoring Illustrated*, 7 September 1907, ads. Howard's 'Road Book' (p. iii) is a good example of the genre: very detailed descriptions of travelling between towns for cyclists, including descriptions of roads and whether fit for cycling, plus distances, hazards and places of interest: Howard, *The Roads of England and Wales*.
[68] Filson Young, *The Happy Motorist*; Matson, *The Modest Man's Motor*; Max Pemberton, *The Amateur Motorist* (London: Hutchinson and Co, 1907).
[69] See Silvester Mazzarella, 'Filson Young (1876–1938): The first media man', at http://richarddnorth.com/archived-sites/filsonyoung/, accessed 18 October 2019.
[70] 'The Road, Yesterday and Today', broadcast 10.30–10.45pm, 23 November 1928. See *Radio Times*, 268, 16 November 1928, p. 45.
[71] Filson Young, *The Happy Motorist*, p. 12.

as an 'old crock', the tours described (and encouraged) in magazines and books were increasingly made easier, less demanding, and more comfortable. Coulbert described 'the new species of "motor tourist"', able to 'see nature in more personal and non-prescriptive ways, as drivers were invested with new liberties to discover landscape for themselves.' She showed the guide-book writers as seeing the motor car 'open up' the countryside 'for the purposes of knowing and exploring one's own country'.[72] Adverts for motor-cycles, motor cars and cycles in the 1900s were keen to promote their proven reliability because such claims could accompany imagery or testimonials suggesting an ease and suitability of the product for touring. Such tours increasingly contrasted with the pioneers' tour undertaken by the first cyclists and motorists.

This new tourist, emboldened by reports in cycling and motoring magazines, clutching his or her guide book and new map (probably now with contours), now expected road signage, hospitality, passable roads, availability of petrol, spares, service, even, perhaps, a motor car that could go the distance entirely in top gear. Much of the literature at the tourist's disposal was geared towards encouraging an expectation of comfort, reliability and predictability while on tour, and was increasingly for a much broader clientele. 'Owen John' pointed out how, 'for persons touring the British Isles in summer there is no need, with a modern car, to assume any fanciful get-ups ... the best and most useful type of garments are those which are in no wise different from those worn by ordinary folk on ordinary occasions'.[73] Major Matson in many ways represented this new breed of driver. In 1903 he related the story of a driver, who had a tyre blowout, colliding with a telegraph pole, and ending up in a ditch with a second blown tyre. That driver simply fitted new tyres and proceeded on his way. This sort of 'sportsman' was something Matson could never be:

> I cannot enjoy these risks, being a family man, and I do not use pneumatic tyres – I would rather go tiger-shooting. Some day I hope to have a really fast car. I am only waiting til all the main roads are properly straightened out, side roads stopped up, all dust and flints removed, and cyclists, pedestrians, horse traffic, and dogs, compelled by law to travel in suitable tubes underground.[74]

In contrast to the gruelling schedules of the pioneers, Filson Young was advising tourists not to worry about putting in miles, since it was now taken as read that the vehicle would be reliable enough to allow this. He suggested that the tourist start early and put in much of the distance before lunch, rather than thinking that covering large distances was necessary.[75] Other non-motoring magazines picked up on the interest in the motor tour, for example, *The Strand*.[76] The age of the pioneer was over.

---

[72] Coulbert, '"The romance of the road"', p. 216.
[73] 'Owen John', 'Introduction', in A. J. Wilson (ed.), *Motor Trips at a Glance: In England, Scotland, Wales, Ireland and France* (London: A. J. Wilson & Co., 1911), p. 15.
[74] Matson, *The Modest Man's Motor*, pp. 34–6.
[75] Filson Young, *The Happy Motorist*, p. 209.
[76] The magazine featured articles on long-distance travel by motor car in, for example, the United States: Frederick A. Talbot, 'Across America by motor-car: Some remarkable and thrilling transcontinental journeys', *The Strand*, 31 (1906), pp. 513–21.

The 'consumer's' tour was created by magazine editors publishing travel articles, by the setting up of new local businesses (agencies, spares, accommodation) or adapting what was already there (blacksmiths taking on motor repairs, selling petrol in larger quantities). Local clubs patronized these on their tours. Going for a picnic, or to a beauty spot, was encouraged by being portrayed as respectable and desirable. A language of ease and carefree affordability was adopted. It has been shown how magazines continually revisited the ongoing costs of running a motor-cycle or motor car; they also offered costs for touring. In *Motorcycling*, a few years later, it was still thought necessary to encourage more motorcyclists into the fold by playing down running costs. 'The man of moderate means may be inclined to think that motorcycling is an expensive pursuit', but this is not the case. 'A man with £3 a week should have no qualms about buying a machine for fear of heavy running expenses. It is probable that he will only use it at week-ends and at holiday times, and in that case a £15 note should easily cover a year's expenses'.[77] Mary Kennard identified the 'small fry' of the Thousand Mile Trial in 1900, pointing out how the light-car drivers knew that the drivers of the bigger cars 'looked down upon us, but we did not look down upon ourselves'. She pointed out that the smaller cars and their drivers endured the same trials. She concluded, 'So, dear public, hurry up, take the plunge, be happy, be courageous, join the band of small fry, who have gained an honourable place in the recent contest, an open inviting arms to you to enter their ranks'.[78] Kennard, then, was picking up on a theme already discussed, of how potential consumers may have kept away from motor traction if they felt they might be exposed as barely able to afford to engage. Her very language betrayed the pecking order among motorists, made more raw by the ability of wealthier motorists being able to tip. For example, *Bicycling News* noted in 1900 how,

> At present the motist is fair game for any amount of swindling and blackmailing [by hoteliers]. [. . .] In the Motor Car Club's trip to Brighton last November, the charge for stabling a two-seated car for the night was no less than 5s . . . this imposition [poor and expensive stabling], I may add, was made at a house after a visit from the Automobile Club, the members of which appear to delight in spoiling ostlers and landlords for other drivers, simply to avoid appearing mean.

The Cyclecar magazine was sensitive to this too, and noted how income determined opportunities for leisure. Some might be able to do a weekend, 250–300 miles, hotel and garaging, while others just afford a there-and-back on a Sunday.[79] However, few would be able to match E. Douglas Fawcett, who in his 'Rambles on a light car' column described his three-month tour of Europe in his 6hp De Dion. 'Let not the light car folk dream for a moment that the joy of touring afar is not for them', he said, suggesting a 6 or 8hp car would be the equal of a 40hp. Touring will facilitate the 'independent, roving gypsy life', as you steal from your Swiss hotel or chalet 'to "purr" along the mountain

---

[77] 'The businessman and the motor-bicycle', *Motorcycling*, 28 June 1910, p. 190.
[78] Mary Kennard, 'The small fry of the trial', *Motor Car Journal*, 19 May 1900, p. 205.
[79] *The Cyclecar*, 7 May 1913, pp. 617–8.

ways in quest of new adventure'.[80] Clearly, being able to afford three months of leisure in one's continental hotel, whether by light car or not, would for many have been quite impossible, for whom short trips, and possibly camping, would have marked the limit of aspiration;[81] the dreams sold by the magazines simply came in all sizes.

Horizons were quickly broadened: opportunities arose for women and couples to tour unchaperoned while the trip abroad was informed and glamorized by a press reporting on the lifestyles of the celebrities of British high society in foreign resorts.[82] Travelogues included the increasingly exotic, such as touring in Persia. Part of this process involved the inexorable elimination of the electric and steam motor car as a touring device in favour of petrol; Mom has argued that there was nothing inherently 'inferior' about the electric vehicle, but that it was ultimately undermined by a combination of factors, including recharging infrastructures and being seen as insufficiently masculine. Ivory and Genus describe, instead, the context of the 'lock-out' of the private electric car. 'Viewed through the lens of symbolic consumption and lockout', they observe, 'we begin to see that, as a consequence of an emerging car culture built around class status and masculinity, the electric car had become a culturally and socially untenable object – even in rural settings.'[83] Increasingly, the perceived ability to tour all day informed consumers' preferences for a petrol economy, with the electric car being seen as a 'woman's car', and one with a small range and requiring lengthy stops to recharge batteries. The steam car was also being seen as inconvenient, in that it needed tens of minutes to come up 'to steam' from cold before it could be driven. And where the steam car was demonstrably superior to the petrol, such as in a hill-climb competition, clubs started to disallow the steam. For example, the Manchester Motor Club unintentionally made publicity when, in its 1905 Glossop hill-climb, it barred entry to the steam cars of a Mr Wilkinson, who took the matter to the *Manchester Chronicle*. 'Autolycus', a committee member of the MMC, wrote to *The Motor* to admit that the participation of steam vehicles would have been 'practically a walkover' against the petrol cars.[84]

Another dimension to widen opportunities to tour came with the acceptability of camping. The hobbyist press had been encouraging interest in the outdoors, in particular camping and, later, caravanning.[85] Cycle-touring had been practised since the 1870s although overnight accommodation tended to be at hotels or inns, to which luggage was often sent on in advance.[86] A photo-article in *The Motor* by F. Horsfield, honorary secretary of the Association of Cycle Campers, explained the rationale behind camping: all too often the hotelier will assume motorists are wealthy, but the 'man of moderate

---

[80] *The Motor*, 14 February 1905, pp. 34–5.
[81] See Horner, 'The emergence of automobility in the United Kingdom', pp. 56–75, at p. 66.
[82] Steward, '"How and where to go"', p. 39.
[83] Gijs Mom, *The Electric Vehicle: Technology and Expectations in the Automobile Age* (Baltimore, MD: Johns Hopkins University Press, 2005); Chris Ivory and Audley Genus, 'Symbolic consumption signification and the "lock-out" of electric cars, 1885–1914', *Business History*, 52:7 (2010), pp. 1107–22, at pp. 1107–8, 1118. See also Michael Brian Schiffer with Tamara C. Butts and Kimberley K. Grimm, *Taking Charge: The Electric Vehicle in America* (Washington, DC: Smithsonian Institution Press, 1994).
[84] *The Motor*, 26 December 1905, p. 603.
[85] For caravanning, see, for example, the feature in *The Motor*, 29 July 1913, pp. 1191–3.
[86] See, for example, Bury and Hillier, *Cycling*, chap. 5, 'Touring'.

means' will be glad to know that one can tour with the car for not much more than the cost of the food. Taking cycle-camping up a step, it would be easy for the motorist to do the same. The article identified a list of necessary equipment: tent, sleeping bag, ground sheet, stove, totalling nine pounds.[87] The addition of motorists into the fold enabled new sales opportunities for outlets such as Gamage's or Dunhill's, already dealing in motor-clothing and componentry. Camping equipment, 'specially designed for transport in motor-cars' was now available.[88] Initially, reports in motoring magazines were of adapting the cabin of a motor car by removing or flattening seats to make a bed, and by draping waterproof sheets as an awning. Camping had cross-class appeal: the language of 'vagabondage' and 'a-gipsying' was adopted,[89] and, on the one hand, its very cheapness was used as its attraction; *The Cyclecar* reported on a fortnight's tour in a second-hand AC Sociable (a popular three-wheeled light car) in 1913, for a total of thirty shillings, or ½d per mile, presumably camping.[90] On the other hand, a wealthier clientele was also drawn in to camping. See, for example, Filson Young's remark, to an audience who could afford not to camp: 'We may wish to be really adventurous and to camp out sometimes'. Recognizing there might be some readers who only had one car (!), he agreed that packing camping gear for two would be very tight, so thoughtfully suggested it would be much better to travel with two cars, sending one ahead. This would be particularly useful for a class of camper who needed changes of clothes depending on the circumstance; he suggested that luggage be sent on in advance 'to any place where appearances have to be studied'.[91] The intrepid camping motorist was also reassured that he would find the locals were friendly. 'Owen John''s trip through the Black Country talked of 'yokels',[92] while a camping weekend in Ireland described how the entire population of the nearby hamlet watched them set up camp before asking, '"What toime was the perfarmance going to begin?"'[93] (The 'yokel' as a bewildered onlooker was a *Punch* staple in its cartoons.)

Stopping at the roadside for picnics and to brew up was encouraged by the magazine writers, who published photographs of happy groups doing just that, often in infeasibly idyllic settings. Articles gave advice on the best equipment and camping grounds,[94] while readers continued to supply details of their own modifications to their vehicles for camping.[95] Developments in caravanning were reported in the magazines, with readers describing self-built trailer caravans.[96]

---

[87] *The Motor*, 5 August 1905, p. 162.
[88] Filson Young, *The Happy Motorist*, pp. 182–3.
[89] *The Autocar*, 25 July 1914. 'Gypsying' was having a weekend getaway with caravan or tent: Coulbert, '"The romance of the road"', p. 202.
[90] *The Cyclecar*, 23 July 1913, p. 230.
[91] Filson Young, *The Happy Motorist*, pp. 182, 185.
[92] *The Autocar*, 1 July 1905, pp. 12–13.
[93] *The Cyclecar*, 10 September 1913, pp. 388–9.
[94] Such as Major E. Harvey Jarvis, 'Camping with a car: with hints as to equipment, camping grounds etc.', *The Autocar*, 25 July 1914, pp. 166–9.
[95] See also Oliver Heal, 'Sir Ambrose Heal and the automobile', *Aspects of Motoring History*, 12 (2016), pp. 100–2, in which a letter in *Country Life* in 1922 describes, with the help of scale drawings, the rendering of the cabin of a Standard motor car into a sleeping compartment.
[96] See Roger Ellesmere, *British Caravans, Vol. 1: Makes Founded before World War II* (Beaworthy: Herridge and Sons, 2012), 'Introduction'. See also the despoliation of the rural idyll by caravanners in 'The caravan craze: Scene in a lonely part of the Highlands', *Punch*, 10 June 1908, p. 431.

Encroachment, blight and the working classes

The encroachment of the cyclist, then 'modest motorist', onto the countryside, taking advantage of camping as a cheap means of extending the tour, was accompanied by a working-class incursion by charabanc. This and the municipal (and private) motor-bus service were for Jeremiah 'a key part of the opening up of the countryside' before the First World War.[97] Law has suggested that

> Long before the arrival of the working-class car owner, poorer members of society achieved a spontaneous, autonomous mobility of their own through the charabanc. This disconnected them from the more controlled and directed forms of public transport that had previously taken the working classes to the seaside such as the train and the steamer.[98]

With the first motorized 'charries'[99] appearing about 1900,[100] by 1914 *The Motor* reported in a special feature on the Manchester area how there had been a 'wonderful rise in the popularity of the motor char-a-bancs in the Lancashire, Cheshire and North Wales area', to 'get them about the country at weekends'. This had 'done much to popularize motoring amongst the "lower orders" . . . in all grades of society motoring as a recreative pastime has progressed during the past year in a manner surpassing all previous periods'.[101] Charabanc users were also keen to get their money's worth: an account in *The Car Illustrated* in 1905 described a motor omnibus trip from London to Brighton and back, which stopped in Brighton for a mere seventy-five minutes before having to return.[102] Charabancs and shared ownership of used motor vehicles continued a trend towards affordable working-class excursions.

The travel writer and social commentator H. V. Morton (1892–1979) was unimpressed. He believed 'this lower class of traveller had neither the education nor the taste to appreciate the country in picturesque terms, contributing to the "vulgarisation of the countryside"'.[103] Morton was not alone here: a common theme for most travel writers was a preoccupation with industrialism, idealizing pastoral landscapes, and class prejudice against working-class tourism.[104] In (vain) hope a charabanc driver F. J.

---

[97] Jeremiah, 'Motoring and the British countryside', p. 240. See also B. Gibson, 'From the charabanc to the gay hostess', in Robert Snape and Daniel Smith (eds), *Recording Leisure Lives: Holidays and Tourism in Twentieth-Century Britain* (Eastbourne: Leisure Studies Association, 2011); Stan Lockwood, *Kaleidoscope of Char-a-bancs and Coaches* (London: Marshall, Harris and Baldwin, n.d. [c. 1980]); J. K. Walton, 'The origins of the modern package tour? British motor-coach tours in Europe, 1930–1970', *Journal of Transport History*, 32:2 (2011), pp. 145–63.

[98] Law, 'Charabancs and social class in 1930s' Britain', pp. 41–57 at p. 42. See also Michael John Law, *The Experience of Suburban Modernity: How Private Transport Changed Interwar London* (Manchester: Manchester University Press, 2014); see also Law, 'Driving to the "Super" Roadhouse', pp. 49–60.

[99] Coulbert uses the word: '"The romance of the road"', p. 215.

[100] Lockwood, *Kaleidoscope of Char-a-bancs and Coaches*, esp. chap., 'The pioneers'.

[101] 'Motoring in the Manchester district: The year's progress in sport and trade', *The Motor*, 6 January 1914, pp. 1171–3.

[102] *The Car Illustrated*, 6 September 1905, p. 79.

[103] Morton, *In Search of England*, cited in Coulbert, '"The romance of the road"', pp. 211–12, f. 38.

[104] Coulbert, '"The romance of the road"', p. 201.

Layzell wrote a letter in 1920 in response to a recent campaign against the charabanc. 'It is to our interest to encourage Tom, Dick and Harry to motor as much as possible', he wrote. 'There is enough class prejudice in the ordinary walks of life – for heaven's sake let us try to keep it out of the motor movement'.[105]

Another perceived blight was the appearance of advertising hoardings, and here commentators recognized that a profound change had overtaken village high streets and their hinterlands in the period to 1914.[106] The putting in place of an infrastructure fit for motor traction was not initially subject to any planning restrictions, and the resulting eyesores led to calls for the protection of the countryside. Motoring magazines picked up on this issue, with *The Motor*, for example, drawing attention to 'irregular and excessively ugly' hoardings, 'obliterat[ing] entirely the view of the country in the background', before going on to take credit for its part in their removal.[107] *Punch* cartoons mocked the noise, dust and chaos that motorists now brought to the high street.[108] Hissey commented on this: 'During our journey [in 1906] we observed many a pleasing bit of village architecture thus spoilt as a picture by glaring plaques of crudely coloured enamelled iron attached to them, setting forth the virtues, real or otherwise, of somebody's soap, some other body's ointments or pills, and the like. One day we may discover the value of the picturesque and take measures against its spoliation, for nothing now is sacred to the enterprising advertiser'.[109]

Speed trapping

One reason for 'E. M. B. P.' to tour abroad was that he perceived a greater freedom there on the open road. 'Having suffered at the hands of the British police for exceeding the speed limit on an open country road early in the morning', he wrote, 'I feel rather afraid to stay in my own country for a holiday.'[110] He was referring to the widespread practice of speed trapping which had been in place since the passing of the 1896 Locomotives on Highways Act and subsequently endorsed by the 1903 Motor Car Act; these Acts set the speed limit on the open road to 12mph and then 20mph, respectively. Most motor vehicles were fully capable of exceeding these speeds.[111] Speeding cyclists, that is, those who 'scorched', had already been at the receiving end of legislation which

---

[105] *The Light Car and Cyclecar*, 3 July 1920, p. 117.
[106] David Jeremiah, 'Filling up: The British experience, 1896-1940', *Journal of Design History*, 8:2 (1995), pp. 97-116, esp. images on pp. 98-9; Jeremiah, 'Motoring and the British countryside', pp. 233-50, at pp. 234, 239.
[107] *The Motor*, 25 March 1913, pp. 351, 361.
[108] For example, 'A quiet Sunday in our village', *Punch*, 6 June 1906, p. 412.
[109] Hissey, *Untravelled England*, pp. 18-19, cited in Coulbert, '"The romance of the road"', pp. 210-11.
[110] *The Autocar*, 15 January 1910, pp. 80-2.
[111] See Clive Emsley, '"Mother, what did policemen do when there weren't any motors?": The law, the police and the regulation of motor traffic in England, 1900-1939', *Historical Journal*, 36:2 (1993), pp. 357-81; Layhourn with Taylor, *The Battle for the Roads of Britain*, esp. pp. 45-54; Plowden, *The Motor Car and Politics in Britain, 1896-1970*, esp. chap. 4, 'An association of burglars'; Keir and Morgan (eds), *Golden Milestone*; Cooke, *This Motoring*.

had defined 'furious driving'[112] and now motorists were finding the penalties were potentially severe. C. Uchter Knox, for example, went on a trip from Hampshire to the Lake District in his 30/40hp Daimler in 1910, and on the road beyond Carlisle he got caught in a speed trap which cost him the maximum fine of £10 plus an endorsement of his licence.[113]

Trapping varied in its severity from one county to another, at the whim of the chief constable and magistracy, and the subject had excited the outrage of motorists from the outset. Magazines began to publish maps identifying the 'worst' roads, *The Autocar* first in 1907, and these were updated weekly[114] with readers' letters advising of any developments; from 1905 in *The Motor* the letters' pages had a 'Police traps' section.[115] All magazines shared a feeling of victimization, and did so even beyond the First World War. Jarrott and William Letts, the agents (over time) for Crossley, Sizaire, de Dietrich, Dodge and Oldsmobile, had from 1905 offered a 'service' on the London to Brighton road employing cyclists to wave 'Jarrott and Letts' flags, warning motorists of speed traps. This sales initiative had played well with motorists, as letters attested.[116] It was natural, then, that on the formation of the Automobile Association (AA) that same year Jarrott would be invited onto its committee, along with Edge. The AA was created to thwart police trapping and with MPs amongst its members, it had high-powered support. It enjoyed mixed success, though, particularly with its policy of employing 'scouts' on bicycles whose job was to salute oncoming motorists, or fail to salute them to indicate a trap ahead. A series of legal test cases were brought to challenge the AA's methods and to establish whether the scouts were in effect perverting the course of justice by preventing the prosecution of an offence, and tensions between the AA and the police continued long after the First World War.

The method for trapping was described in most motoring magazines,[117] although the method was doubtless familiar to most cyclists and motorists at the time. Two plain-clothes police officers hid along a suitable stretch of road with a known distance, ideally a furlong, between them. As the selected motorist passed the first officer, he signalled to the next officer, who started his stopwatch. If the motorist passed the second officer within a known time, it would mean he had been speeding, and so that officer would then signal to a third officer, in uniform, who would step into the road to flag down the motorist. Having given his name and address to the officer, the motorist would in due course receive his summons, and, if the letter writers are to be believed, he could expect little mercy in court. The injustice was spelt out in a commentary in *The Lightcar and Cyclecar* in 1913:

> Some years ago it was considered evidence of dangerous scorching if a driver had his licence endorsed; but owing to the great increase in the number of police

---

[112] See Reid, *Roads Were Not Built for Cars*, chap. 5, esp. p. 87.
[113] *The Motor*, supplement to 27 December 1910, n.p.
[114] See for instance *The Motor*, 14 April 1914, p. 512, inc. diagrams.
[115] *The Motor*, from 27 June 1905.
[116] *The Autocar*, 29 April 1905, p. 601.
[117] See, for example, 'A police trap: How it is worked', *Motoring Illustrated*, 23 August 1902, p. 294.

proceedings, it is now considered merely hard luck and no disgrace if one is caught in a police trap.[118]

Certainly, the apparent arbitrariness of trapping played into the motorists' hands. In a cartoon of a broken-down motor car being pushed in the rain, a youth looking on offered, 'I feel it my duty to warn you that there's a police trap round the corner'.[119] *Motoring Illustrated*, commenting in 1902 when there appeared to be many speed traps to contend with said: 'the feature of the run [to Winchester] was the ridiculous action of the police, which had, however, the good result that it showed clearly that the sympathies of the non-motoring public are with the motorist and against the police.' The practice was sufficiently widespread to catch the attention of the non-motoring press, for example, *The Sphere* in 1908. In his 'The motorist's and golfer's notebook', Max Pemberton had much to say on the subject, deciding that on the Old Kent Road from London to Maidstone, even the 'most reasonable driver' must have objected to the '"furlong" tests on perfectly straight stretches and to their annoyances'. As for the Canterbury-to-Sandwich road, this was 'wonderful' but noted for its heavy police presence. Their traps were 'sometimes fair, sometimes the reverse', but while 'there is every excuse for a "control" in the narrow winding streets of Surrey, [there was] none whatever for the traps out towards Upminster and upon the Ramsgate and Margate roads'.[120] *Punch*, of course, lapped it all up. Typical was its series of cartoons 'Waiting for bigger game', where a policeman allowed by a speeding horse and trap, then a scorching cyclist, finally apprehending a passing broken-down motor car being towed by a horse.[121] Much amusement was derived in creating a stereotype of the dim police officer. A cartoon showed an officer who found himself stuck up a tree. He was nevertheless quite visible to passing motorists, one of whom stopped and asked: 'Can I be of any assistance?'[122] And in its 'Motoring phenomena and how to read the signs' full-page series of cartoons, *Punch* illustrated possible disguises for an officer that the motorist might look out for, for example, a 'yokel' with a stopwatch, and other clues such as overly large footprints leading to a roadside bush.[123]

Thus, while motorists as a collective felt hard done by, and a culture of justification endorsed the anti-trapping tactics of the magazines and clubs, there were other tensions in the mix, such as working-class police officers apprehending middle-class motorists, a class that hitherto was thought to be hardly capable of crime. Nor would it help to imagine a collective of bloody-minded magistrates and police officers, all intent on persecution and refusing to face up to the inevitable mass use of the road by motorists, when other factors were in play such as cyclists who continued to scorch, and the presence of an increasingly vocal and highly connected motoring lobby which sought to maximize its own interests.

[118] 'How a police trap is worked', *The Lightcar and Cyclecar*, 17 November 1913, p. 695.
[119] *Punch*, 3 July 1907, p. 9.
[120] *The Sphere*, 19 September 1908, pp. 32–4.
[121] 'Waiting for bigger game', *Punch*, 22 October 1902, p. 285.
[122] *Punch*, 14 April 1909, p. 267.
[123] *Punch*, 17 January 1906, p. 43.

Conclusions

Charles Jarrott called them the 'old brigade', the pioneers for whom cycling, then motoring, was truly an adventure, a sport. To labour underneath a stricken motor vehicle, or to endure the provocations and missiles of bystanders as a cyclist, were all part of their wider late-nineteenth-century experience of travel. Roadside accommodation was generally grim, and the roads themselves quite unsuited to the speeds and distances that the pioneers determined to achieve. It might have been – it *was* – miserable at the time, but the first cycling and motoring tourists valued their experience, marvelling in its novelty and its excitement. Theirs was the 'adventure machine'. This was the time when the modern and the rustic were first juxtaposed. Motoring-magazine editors encouraged a writing and pictorial genre where the cycle, then the motor car, 'opened up' the countryside, and for the first time it was possible to get to know and to explore one's own country all in the space of a weekend. City dwellers could now 'get back to England's Arcadian past' and 'escape the regimen of industrialism'.[124] The 'open road', the beauty spot, the vantage point, the rustic scene, all could now be reached and gazed upon in an icon of modernity, a mastery of man over machine. The all-new thrill of speed on the way embodied power, progress and independence.

It has been suggested here that within a short space of time, probably by about 1905, the nature of the tour, and of the tourist, changed. The cycling diarist G. H. Smith, writing in the 1930s of the 1880s and 1890s, remarked, 'one could ride on a solid[-tyred bicycle] for years and not be conscious of any discomfort, but once you rode a pneumatic or cushion[-tyred bicycle] for a few weeks, it was absolute agony to go any distance on a solid afterwards'.[125] Such was the experience of the later tourist, the 'consumer' tourist. The infrastructure to permit their wider consumption of the countryside had started to develop: the availability of petrol, the readiness of the local blacksmith to carry out repairs. The 'uglification' of the countryside was to be a by-product of the interwar period, but uncontrolled advertising and signage, a blight for so many observers, gave a hint of what was to be. Despoliation, then, was inevitable. Entrepreneurs saw opportunities in sending the working classes on excursions by charabanc. The impoverished motor-cyclist or motor car owner was persuaded that he could go on tour, by doing a there-and-back on the same day, or by camping overnight. Specialist shops, already providing the accessories and clothing that the driver did not even know he had to have, were now diversifying to meet the demand. Car owners were quick to adapt their vehicles so that they might sleep or shelter in them overnight.

This new tourist expected reliability and a trouble-free journey. He now demanded guide books telling him which hotels were acceptable, and maps with contours which warned him of 'dangerous' roads. The magazine editor had created the dream, and the entrepreneur had made it happen: the availability of more and cheaper motor-cycles and motor cars, the petrol and spares, the lobbying for road 'improvements', signage and accommodation.

[124] Coulbert, '"The romance of the road"', p. 216.
[125] Smith, *Some Notes about the Anerley B.C.*, p. 14.

Travel, for its own sake, was nothing new. The culture of travel already existed, with tourists doing the same circuits, at home and abroad. Now they were doing so covering greater distances in much less time by bicycle or motor traction, taking with them cameras and sketch-books to record it all. They increasingly relied on the guidance of the travel writer, who could take all the stress out of finding the best hotels or bypassing the less rustic towns. This collective of knowledge required constant updating, requiring a new edition to be bought each year. All the time it was plain that the idealized imagery, of the lone car parked overlooking a scenic view, its beautiful occupants enjoying a carefree picnic, was according less and less with the reality of the experience of more and more motorists using the roads at the same time. How they imagined their access to, and use of, the 'open road' would change, is the subject of the next chapter.

7

Futures

A front cover of *Motorcycling* magazine in 1909 featured a drawing of a motor-cycle tearing along the road of the future. 'Speed and comfort on a seven-cylinder 50hp single-track sociable', the caption read. This was no ordinary motor-cycle, with its two-abreast seating and outrigger wheels[1] but, rather, a vision of what the motor-cycle might become. It referenced the latest technology: having a seven-cylinder engine suggested it was a radial aero engine (and therefore very much the latest thing), while as a 'sociable' it was where the illustrator imagined the motor-cycle-and-sidecar arrangement might go next. *Motorcycling* was simply following trends, but now assumed motor traction was here to stay, and was sure to bring evermore speed and comfort. (It didn't think it was going to bring much more reliability, though, because it, like most commentators, had agreed by about 1910 that the internal-combustion engine was already practically perfect.)

This was one of many contrasting futures for motor traction. On the one side, they tended to fall into the positive and optimistic, peddled by the entrepreneurs and magazines. For them the future was going to be rosy. Roads would be clear, wide and straight, just like the French ones. The dust 'menace' would be eliminated. Fiction, such as short stories in the weekly middle-class press (*The Strand*, for example) often featured the motor car as omnipotent or invincible (for example, racing a train – and usually beating it).[2] A motor car had, after all, already exceeded 100km/h even before the turn of the century (the *Jamais Contente* driven by Camille Jenatzy in 1899) and as soon as 1907, S. F. Edge was driving his Napier for a sales stunt at an average speed of over 100km/h for twenty-four hours non-stop. These feats, and the more down-to-earth ones such as trials and hill-climbs, added credibility to motor traction as an every-day activity and enabled illustrations of seven-cylinder motor-cycles to appear to signal what might be. The public imagination was being piqued by reports of the reality (the land-speed record, the clearly improving reliability), which in turn added credence to the artistic visions of the future.

Another future, though, was presented by a rather reactionary non-motoring press. *Punch* cartoons, by about 1905, no longer featured broken-down and uncontrollable motor vehicles with the 'yokel' looking on but, instead, village high streets hopelessly clogged by motor vehicles, with dust, noise, advertising signage and aggravation the

---

[1] *Motorcycling*, 6 December 1909, cover.
[2] See also Jeremiah, *Representations of British Motoring*, pp. 18–26; See Arabella Kenealy, 'Twixt cup and lip', *Motoring Illustrated*, 3 May 1902, pp. 212–4.

new reality.³ *Punch* was also one of many popular magazines to publish cartoons of rather more incredible, tongue-in-cheek visions. For example, and anticipating cars of the 1920s, a cartoon featured a car driven by a propeller which could be adjusted to cause the car to fly.⁴ (*Punch* also published cartoons of flying policemen.) While these cartoons had no basis in engineering, and played on ignorance or fantasy, they were feeding off aviation developments then in the news. This chapter, then, will navigate a path between these two visions. It will take a broad look at the anticipated new possibilities for motorists (such as private aviation), and who was expected to indulge in them. It will look at how the light car, of acceptable quality and breaching the psychological barrier of £100, was always just 'around the corner'. It will consider how the gap was filled, if in reality only by entrepreneurial rhetoric, by the arrival of the 'new motoring' cyclecar movement from about 1910 (predating the mass participation brought about by the Model T Ford and the 'Bullnose' Morris Oxford). It will look at how it was imagined that the road network would change to accommodate all these new motorists, but also how the horse persisted in defiance of expectations of a motorized future.

New possibilities beyond motoring

Ballooning was fashionable in the first decade of the twentieth century and quickly came under the remit of the Aero Club, founded in 1901 by, amongst others, pioneer motorist and adventurer Frank Hedges Butler. The Club put on publicly viewable ascents weekly, from the Crystal Palace grounds and elsewhere, although only members (and even they, after having paid a guinea) could participate; this very visibility kept the sport in the public eye and imagination. The celebrated aviator Alberto Santos-Dumont (1873–1932) was interviewed for the new *Motoring Illustrated* magazine in 1902. 'I have no hesitation whatsoever', he said, 'that the voyage from London to New York and back by balloon will be perfectly feasible in the course of a few years'. Santos-Dumont, the 'sky-*chauffeur*', had already achieved 40mph in a balloon, and was working on making balloons more manoeuvrable, able to circumnavigate obstacles such as towers.⁵ Readers of *Motoring Illustrated*, or, indeed, of any of the motoring hobbyist press, would have seen no contradiction in the reporting of developments in the aviation world, just as they were interested to read about motor-boating. And once such articles in turn generated a tangible and sufficient readership, new dedicated magazines were spun off; for Dangerfield's Temple Press, *Motorcycling and Motoring* became *The Motor*, which in turn spawned *The Motor Boat* (1904, ed. George F. Sharp); *Commercial Motor* (1905, ed. Edward Shrapnell-Smith); the revival of *Motorcycling* (1909, ed. W.G. McMinnies); and *The Cyclecar* (1912, ed. A.C. Armstrong).

Flight had similarly captured the imagination of many sportsmen and entrepreneurs who also had their interests in motor traction. *Flight* magazine

---

[3] For example, 'A quiet Sunday in our village', *Punch*, 6 June 1906, p. 412.
[4] *The Motor*, 17 January 1905, p. 662.
[5] *Motoring Illustrated*, 15 March 1902, p. 35.

appeared first in 1909, weekly for a penny (published by F. King and Co. as a spin-off from *Automotor Journal*). Temple Press followed later that year with *The Aero*, again weekly for a penny (ed. Charles Grey), and also published the annual *Aero Manual*. But aviation was also suitably exclusive by way of sheer expense and need for club membership. Maurice Egerton of Tatton Hall in Cheshire described himself as an 'aviator' in the 1911 census and serves well here as an example of a sportsman fully immersed in motoring, aviation and adventure. He had a succession of large Darracq motor cars, probably acquired through his connection with the motor agent and aviator Toby Rawlinson (1867–1934).[6] Meanwhile, the motor salesman and playboy Charles Rolls was a keen balloonist, who moved to powered flight at the first opportunity; he made the first flight in the United Kingdom in a powered dirigible in 1907.[7] Edge, in contrast, did not appear to have embraced aviation with the same enthusiasm. He was known, however, to have made at least one flight by balloon. In 1902 his maiden flight was in *Shropshire*, in Sir Vincent Kennett-Barrington (1844–1903)'s balloon, accompanied by the balloonist Percival Spencer (1864–1913).[8] Ascending from the Ranelagh Club grounds, they rose to 35,000 feet, and descended near Bicester.[9] Edge probably did not share Santos-Dumont's enthusiasm for ballooning – he made no grand pronouncement at the time – but he did see commercial potential in powered flight a few years later. By 1907 he was on the provisional committee of the Association for the Promotion of Flight and represented the Aero Club and Aeronautical Society.[10] A little later still, at the 1909 annual dinner of the agents' section of the SMMT, he aired his view that a new 'airway' from London to Brighton should be constructed, upon the initiative of the motor-car industry. It should begin at Purley Corner (South London), be 200 feet wide, and could be rented by enthusiasts. All you would have to do, he reasoned, was to clear the trees. This would allow enthusiasts to fly over at weekends. 'I could sell a hundred aeroplanes tomorrow,' he said. 'The new sport would at once have a commercial outlook. Already it is clear that it will be cheaper for one or two people to travel by aeroplane than by motor-car. Motor-car makers must look seriously to the commercial side of aviation.'[11]

Entrepreneurs, then, saw real potential in aviation being the next big thing. Hedges Butler inaugurated his Challenge Cup to award prize money for flying feats and in 1906 Alfred Harmsworth (by then Lord Northcliffe) offered a prize

---

[6] For Rawlinson see Dyson, 'The adventurous life of Sir Alfred', pp. 41–58.
[7] For surveys of aviation development, see Hugh Driver, *The Birth of Military Aviation: Britain, 1903–1914* (Woodbridge: Boydell Press, 1997), chaps. 1 & 2; Dorthe Gert Simonsen, 'Accelerating modernity: Time-space compression in the wake of the aeroplane', *Journal of Transport History*, 26:2 (2005), pp. 98–117; Lucy C. S. Budd, 'Selling the air age: Aviation advertisements and the promotion of civil flying in Britain, 1911–14', *Journal of Transport History*, 32:2 (2011), pp. 125–44; D. McCullough, *The Wright Brothers: The Dramatic Story Behind the Legacy* (London: Simon & Shuster, 2015).
[8] *Illustrated London News*, 7 June 1902.
[9] Duncan, *The World On Wheels*, ii, p. 840; *Automotor and Horseless Vehicle Journal*, 14 June 1902, p. 231.
[10] *The Times*, 24 January 1907.
[11] *Daily Mail*, 17 November 1907.

of 10,000 pounds for anyone who could fly from London to Manchester within twenty-four hours and with only one stopover. (*Punch* picked up on the apparent absurdity of such a challenge, reckoning Harmsworth's money was safe). But cartoons in *Punch* also imagined that the aeroplane would be 'normal' in five or ten years, and that it would replace, or at least complement, the motor car as everyday transport. Writers such as Major Matson published guides on aviation for the beginner (in his case, the budding back-yard aviator could build his own for four pounds). He reckoned that while 'the public are agog for the spectacle' of aviation shows, those shows would soon cease to make money, once people tired of craning their necks to look upwards, and soon aeroplanes would become as common as bicycles.[12]

The cycle- and motor-car manufacturer Humber announced in 1909 that it would produce cheap aeroplanes, and signalled the flotation of a company formed to build large dirigible airships. The board included Edge and, amongst others, Robert Baden-Powell (1857–1941), and the *Daily Mail* trumpeted how 'the company aspires in its name and object to national service'.[13] Furthermore, in 1913 Edge was one of many industrialists and public figures who presented a 'memorial' to lobby Prime Minister H. H. Asquith as president of the Committee of Imperial Defence, requesting at least one million pounds 'as shall be sufficient to place the British Air Services on a thoroughly satisfactory basis', to overcome the lack of aeroplanes and airships.[14] Meanwhile, Wolseley, having been taken over by Vickers, Sons & Maxim Ltd in 1901, had interests before the First World War in submarine engines and aeroplanes, plus an airship for the Royal Navy.[15]

Harmsworth's prize money was eventually claimed in 1910 when the Frenchman Louis Paulhan (1883–1963) pipped Claude Grahame-White (1879–1959, brother of Montague) to the post, landing in a field in south Manchester. As a measure of how flight had captured the public imagination, the landing was witnessed by a crowd of 3,000–4,000.[16] While much was said in praise of the Frenchman, Harmsworth's newspaper the *Daily Mail* painted the runner-up as a hero: 'he had made a grandly British struggle for victory, and has accepted his disappointment with grandly British generosity. No wonder people wanted to shake hands with this gallant boy'.[17] As in the case of motor racing, the British loved their sporting heroes.

Motor-boating had a similar cachet, with articles on this subject in the motoring magazines. The racing motor-boat did not exist until engine-manufacturers like Daimler put internal-combustion engines in boats from the 1880s, followed by others such as Napier from 1898.[18] Edge entered Napier-engined motor-boats in prestigious regattas and races such as in Deauville in 1903, Monaco in 1904 and the Mediterranean

[12] Major Matson, 'Flying for all: The modest man's monoplane', *Badminton Magazine*, 31:180 (July 1910), pp. 33–47.
[13] *Daily Mail*, 4 October 1909.
[14] *The Times*, 17 March 1913.
[15] Clausager, *Wolseley*, esp. chap. 2, 'Enter Vickers', and pp. 91–2.
[16] *Manchester Guardian*, 29 April 1910.
[17] 'Desperate chase. Mr Grahame-White in the dark. Pluckiest flight in the contest', *Daily Mail*, 29 April 1910.
[18] Beaulieu and Burgess-Wise, *A Daimler Century*, chap. 1; Vessey, *By Precision into Power*, p. 47.

Cup (Algiers to Toulon) in 1905. There was a clear public interest in motor-boat racing that Edge created, encouraged and tapped into. For example, in 1904 *The Times* published details of the British International Cup for Motor-Boats, and that same year the Crystal Palace Automobile Show was including speedboats amongst the exhibits. The Olympia Automobile Show in 1905 also featured motor-boats, and for *The Times* correspondent, it was a 'real pleasure' to say that the Napier-engined boats exhibited by S. F. Edge (Ltd) 'are nearly as good as they can be'.[19] It has already been seen that Dangerfield launched *The Motor Boat* in 1904,[20] while interest in the cross-channel swimming attempt by Samuel Wilson Greasley (1867–1926), John Haggerty (1862–1939) and one-time racing-cyclist Montague Holbein (1861–1944) in 1904 was an opportunity for Edge to be seen to make his speedboat *Napier Minor* available.[21]

As *The Motor* did to encourage a wider motoring, *The Motor Boat* ran articles suggesting that for only a few hundred pounds, motor-boat ownership beckoned. Even the Automobile Club was undertaking reliability trials for motor-boats, in July 1904. As *Punch* had published cartoons of the skies full of aeroplanes, it now published cartoons of marinas full of motor-boats. Picking up on a *Daily Telegraph* report on a meeting of prominent yachtsmen, who were to found the Marine Motor Association, *Punch* published a cartoon of where this might lead. With the caption, 'our anticipatory artist has a vision of an endless vista of pleasant marine-motor week-ends', it featured floating motor cars whose drivers vied for space, while a floating police car homed in.[22] Meanwhile, by 1904 Edge was making Napier motor-boats available for navy manoeuvres.[23]

Futures: for whom?

However, in the anticipated futures of the motor car and the motor-cycle, it was clear that there were differing visions of who exactly should be included. In 1906, Montagu invited noted pioneer motorists to contribute to his volume on the history of automobilism. Montagu contributed a brief note on where he thought automobilism would go from there. It would, he speculated, certainly replace nearly every other kind of traction, with the tram wiped out. It would affect the value of land, towns and houses. Railway companies would benefit or lose, according to the enterprise of their management. Many readers would live to see the railway companies digging up some of their tracks and converting it to motor traffic, while large towns would have arterial roads and dustless motorways built between the cities. *Punch* in 1903 had its own vision: 'The peaceful English lane of the future', featured a chaotic backdrop of traffic, airship-boat trips to the Zambesi for five shillings, petrol available from troughs, a pub called the 'Airship', and a circus with an attraction 'The only non-extinct horse'.[24]

[19] *The Times*, 14 February 1905.
[20] Announced in *The Motor*, 3 May 1904, p. 344.
[21] *Daily Mail*, 13 August 1904.
[22] *Punch*, 26 November 1902, p. 377.
[23] *Daily Mail*, 30 August 1904.
[24] *Punch*, Alamanac for 1903.

Indeed, Montagu anticipated a future where, unlike the present day, there would be little noise, no smell and dustless roads, as 'no bacteria will breed in fermenting horse manure, and the watercart will be unknown'. Farming land would change use; oats would disappear as they would not be needed to feed horses.[25] Edge speculated that the country could draw on its shale-oil reserves to remove the dependency on foreign supplies, and David Salomons had made the link between the demise of the horse (meaning oats were no longer needed) and the possibility of using the redundant farmland for growing alcohol-yielding crops in bulk: 'we all look forward to the day when motors will drink alcohol in the place of mankind [. . .] with the advantage that it can be entirely produced within the boundaries of our island'.[26] These messages offered hope for an island-nation in an increasingly harsh global market.

There would be a new kind of internationalization – Europe in a few years would become, for the motorist, one vast holiday area.[27] For Montagu, a one-time Conservative MP and editor of the art-paper motoring magazine *The Car Illustrated*, the motor car would ultimately have wide cross-class use; writing in 1906, he was commenting with some benefit of perspective and could see just that already happening around him.

Many commentators writing earlier, while not enjoying the advantage of perspective, either had tunnel vision, or more likely, presented a future which was intended to appeal to, and be exclusive to, their privileged readers. For example, *Motoring Illustrated*'s 1902 feature on the motorist Mrs T. B. Browne reported on how she used her Renault for 'shopping, calls, and delightful little runs to pretty country spots within reach of Kensington'.[28] Harmsworth's edited collection for the Badminton Library, *Motors and Motor Driving*, appearing the same year, was also clear in how narrow the appeal of automobilism would be. Montagu's chapter on the future of the motor car imagined it could be used to fetch visiting guests from the local railway station, and once at the country estate, to ferry them and any staff to the shoot. Other uses were alluded to – local deliveries and so on – but the main purpose of the motor car was plain; it was for the convenience and pleasure of the leisured classes.[29] Mrs Roland Browne found her Lanchester motor car in 1902 'invaluable' for 'town work, social calls, shopping and the theatres'.[30] Major Matson, writing in 1903, described the 'inestimable' value of the motor car, giving the example of a (desperately ill, and clearly very important) patient whose motor car was used to collect doctors, three nurses with two relatives, some 'brave and resourceful women', ice, an operating surgeon and an anaesthetist. (His life was saved.) Beyond that, though, even Matson, a champion of the modest motorist, could think only of 'shopping, social duties, or even a blow of sea air'.[31] Furthermore, Jarrott saw the motor car in terms of privilege and leisure: to the field sportsman, he said, inaccessible spots were within reach; to the shooting man, the moor was within

---

[25] Montagu, *The First Ten Years of Automobilism, 1896–1906*, pp. 123–4.
[26] David Salomons in Montagu, *The First Ten Years of Automobilism*, p. 11.
[27] [Lord] Montagu in Montagu, *The First Ten Years of Automobilism*, pp. 123–4.
[28] *Motoring Illustrated*, 5 April 1902, p. 105.
[29] Scott-Montagu, 'The utility of motor vehicles'.
[30] 'Motorists and their cars: Mrs Roland Browne on her 10-horse Lanchester', *Motoring Illustrated*, 26 April 1902, p. 183.
[31] Matson, *The Modest Man's Motor*, pp. 48–50.

reach; to the hunting man it would get you home comfortably after a hard day; to the golfer it would put links within reach. 'It is the latest form of rapidity of the twentieth century, and in any and every way it makes our modern life easier and more pleasant.' He also said that to the businessman for whom time is money, it was a money saver, but one suspects he was only thinking of those in the boardroom.[32]

This was all the audience needed to know. A clear rationale was being offered for their first purchase of a motor vehicle. Their social potential would increase: horse-keepers, said *The Motor*, had an effective radius of twelve miles, but for a motor-car owner, it would be thirty. Country property could only rise in value; in ten years there would not be a horse left in London.[33] But these accounts were part of a process of normalization of motor traction, just as had been necessary with cycling ten or so years earlier (Balfour, the first prime minister to possess a motor car, had also been a keen cyclist, and shocked former prime minister William Gladstone in 1896 by arriving at Hawarden Castle on a bicycle, having pedalled from the station).[34] Imagining such a future for a wealthy and leisured elite also assumed little change in the social system in place, of differentiation by class. This was underscored by their writing of their own histories. This meant that, for example, it was necessary to record examples of a compliant working-class population vocally endorsing their social superiors in their motor cars. Here, the *Motor Car Journal* reported the reception of the Thousand Mile Trial gaggle of motor cars, on the day in 1900 they drove from Edinburgh to Newcastle:

> All the population of Tranent and Gladsmuir turned out in welcome. And right hearty the welcome was, everywhere. The elder people saluted pleasantly, even the stolid fisherman, taking the 'cutty' pipe from his mouth to wave it in token of his goodwill. As for the children, they went fairly wild with joy, cheering with a shrill treble that was the true keynote of delight.[35]

As a quid pro quo, it was assumed that an 'etiquette' would be established for motor driving, where the wealthy motorist should show 'great consideration for other wayfarers', as '[t]he public is still far from being educated as to the wonderful facility with which these carriages can be controlled';[36] paternalism was very much alive in that pre-First World War world.

The 'etiquette' was necessary because while the pedestrian was increasingly imperilled by motor traffic, a narrative was being constructed to deflect blame from the motorist. In 1910, Henry Sturmey wrote about a motorist of 'considerable experience' who in two weeks had knocked over and killed two children and had other 'narrow shaves'. This motorist had, apparently, an 'extremely quiet car' and cars generally are now so quiet that they are dangerous, because people can't hear them: 'With the exception of the bicycle, the motorcar – once anathematized on account of its noise – is today

---

[32] Jarrott, *Ten Years of Motors and Motor Racing*, p. 282.
[33] *The Motor*, 11 March 1903, p. 96.
[34] R. J. Q. Adams, *Balfour: The Last Grandee* (London: John Murray, 2007), p. 37.
[35] *Motor Car Journal*, 12 May 1900, cited in Bennett, *Thousand Mile Trial*, pp. 228–9.
[36] Alfred Harmsworth, 'The etiquette of the motor car', *Harmsworth London Magazine*, 10 (1903), p. 598.

Figure 7.1 *Punch* magazine quickly changed the way it depicted motoring, moving to portray the motorist and motor car as encroaching and menacing on what had been, only a few years earlier, motor-free high streets. This cartoon is by Harold Millar: *Punch*, 6 June 1906, p. 412. Image: Richard Roberts Archive.

the quietest thing on wheels.'[37] The motoring press was consistent in condemning the dangerous driver, and agreeing that motorists should be more skilled. But it was equally clear that pedestrians had to be more responsible.[38] It was not just the motoring lobby that colluded in a debate to shift responsibility onto the pedestrian: R. P. Hearne wrote in *The Strand* in 1912 that with the huge increase in motorized traffic, 1,557 were killed and 35,120 injured in 1911, and that the numbers were increasing yearly. This often led to 'a bigoted attack upon "murderous motors"', but restricting the free circulation of motors was not the answer, he thought. Slow traffic was not necessarily safe traffic, he said. Ten years ago, there were virtually no motors, but more than one hundred people were killed in London by horse traffic. Now, with 50,000 horses removed, 122 had been killed by horse traffic. The horsed van seldom had brakes, its vision was obstructed by its load or hood, and it was ridden by an unskilled boy who had no licence to revoke. Recognizing the eclipse of horse-drawn traffic as a widely used mode, the magazine looked to the likelihood of much faster motorized traffic, and in this reality offered techniques for the pedestrian to safely cross the road. While offering advice for safer driving, it saw the roads as belonging to the motorized vehicles.[39]

There were, of course, other visions for the uses of automobility beyond those of the leisured classes. Motor traction had many imagined futures, all simultaneously.

---

[37] Henry Sturmey, 'My way of thinking', *The Motor*, 20 December 1910, pp. 847–8.
[38] 'The motorist and the pedestrian', *The Motor*, 28 October 1913, p. 568.
[39] R. P. Hearne, 'The perils of the pedestrian: How to minimize road-traffic accidents', *The Strand*, 44 (1912), pp. 390–6.

Frederick Simms (1863–1944)[40] and Salomons were associated with the setting up of the Self-Propelled Traffic Association in 1895 with a broad remit to develop commercial motor traction, and a branch of the Association in Liverpool set up a trial for motorized heavy traffic in 1896, and again in 1898. The Liverpool SPTA could see the potential in developing light and heavy motor traction for localized commercial purposes, and was active in running 'realistic' tests.[41] Motor traction was also being adopted by businessmen. Sir John Macdonald, for example, observed in 1906 that 'if anyone is still under the delusion that the autocar is a plaything of the luxurious, let him go to the Thames Embankment any morning between 9.30 and 10 o'clock. He will see with his own eyes that there are more autocars than private [horse] carriages conveying businessmen to their daily work in the City'.[42]

Commercial possibilities formed one vision by the motoring press (indeed, *Commercial Motor* was spun off *The Motor* in 1905), but as themes for articles they were subordinate to those of leisure. Instead, and typically, motoring magazine features might include the use of motor vehicles adapted for service in the community, such as adapted motor-cycles for the fire service: in 1910 a 5hp V. S. motor-cycle with sidecar was photographed put to use as a tender.[43] Other realistic possibilities included the use of a tricar for outlying postal deliveries,[44] in which a clear advantage in speed and range could be imagined over the postman on foot or even a bicycle. Some of the more bizarre speculations included the motor-cycle-based non-stop postal service, with the rider 'catching' a suspended mail bag just like mail trains did, and another with a motor-cycle scooping up petrol as it sped past a trough of fuel.[45] (This might have meant the motor-cycle could continue indefinitely, but it did not appear anyone had considered the rider's ability to do the same.) By 1914, in a special feature on Manchester's 'progress' in motor traction and sport and trade, *The Motor* found, 'The commercial and business vehicle, too, has found fresh adherents, and its adoption is becoming more pronounced year by year. No business house can any longer afford to ignore the motor vehicle as a means of rapid transport from place to place of light or heavy goods'.[46] Even so, while there were half-a-million motor-goods vehicles in 1939, there had been at least that number of horse-drawn vans/carts in the haulage business in the 1880s.[47]

The possibilities of war and violence were a significant part of the late-Victorian and Edwardian culture. Reports appeared on how the motor-cycle or motor car

---

[40] Simms was a motoring pioneer and entrepreneur, fluent in German and connected with the Daimler companies. He is also noted for his role in the foundation of the Automobile Club (1897) and the SMMT (1902), having accompanied Evelyn Ellis on (probably) the first recorded motor journey in England in 1895: see Tom Clarke, 'The first motor-car journey in England: A driving controversy', *Aspects of Motoring History*, 12 (2016), pp. 67–81.
[41] See Brendon, *The Motoring Century*, pp. 24–30; and Butt, 'The diffusion of the automobile in the North-West of England, 1896–1939', chap. 2.
[42] [Lord] Montagu in Montagu, *The First Ten Years of Automobilism*, p. 14.
[43] 'Motorcycling first-aid fire appliances', *Motorcycling*, 14 February 1910, pp. 344–5.
[44] See the photograph of a postal tricar in Kent: *Motorcycling*, 14 February 1910, p. 354; of an Eagle GPO delivery van: *The Motor*, 1 April 1903, p. 166.
[45] *Motorcycling*, 14 March 1910, p. 467.
[46] *The Motor*, 6 January 1914, pp. 1171–3.
[47] Thompson, 'Nineteenth-century horse sense', p. 63.

might be adapted for use in the theatre of war as a practical alternative to the horse or bicycle. The motor-cycle had clear advantages in the field, 'for ensuring the rapid transmission of intelligence', being able to cover 100–150 miles per day, and anticipating 'bursting through an outpost line or dodging an enemy'. With war widely seen to be 'coming', 'the motor-bicycle as a weapon of war is gaining way', *Motorcycling* suggested. As a method of delivering intelligence, 'it has no equal'. The magazine featured an article, 'Motor-bicycles in the manoeuvres: An ingenious motor-tricar gun-carriage'. Adapted AC Sociables were shown carrying reserve ammunition and spare parts and fitted with machine-gun mounts,[48] while a full-page artist's drawing featured a gung-ho motor-cyclist, brandishing a smoking revolver and escaping the enemy who was in vain pursuit in a motor car.[49] A more down-to-earth case study was of the 25th County of London Cyclist Regiment, featured camping near Lewes, whose company had already brought along their own motor-cycles (one of which was a home-built contraption, which was not unusual for the pre-war period).

The tank was an innovation of the latter part of the First World War, but had a precursor in the Simms Motor Scout, first announced in 1899. Fully armoured, and built by Vickers, Sons and Maxim, with Maxim machine guns, this quadricycle did not get beyond the prototype stage.[50] The war-car was apparently able to negotiate rough ground, but *The Motor* published a cartoon in which an example, triumphant in the fictitious 1906 invasion, was crippled by a small boy who approached the car with a pin and punctured a visible tyre.[51] More pertinently, the potential of armoured motor traction was not lost on employers and police forces in strike-breaking: the cover of *The Car Illustrated* featured a photo of 'Liverpool policemen in a protected motor-wagon',[52] at the time of the Liverpool general transport strike in 1911.

Entrepreneurs were keen to promote agricultural possibilities, in particular the tractor. In 1902 Daniel Albone (1860–1906)'s 'Ivel' 'agricultural motor' had the commercial support of Edge, Jarrott and even Dorothy Levitt (identified as a 'spinster') as directors of Albone's company. *The Motor* contrasted a photograph of a bullock-pulled plough with the 10hp Ivel pulling the same, all for 300 pounds.[53] A feature in *The Car Illustrated* in 1914 on the Ivel told of how it could do the work of nine horses in less time, at a lower cost, and more efficiently. It could also be used as a stationary

---

[48] 'Motor-bicycles in the manoeuvres', *Motorcycling*, 2 August 1910, pp. 308–9.
[49] 'Running the gauntlet', *Motorcycling*, 21 March 1910, pp. 480–1.
[50] Montagu and Burgess-Wise, *A Daimler Century*, pp. 59–61.
[51] 'A new motor war-car, The Simms, shown at Crystal Palace last week', *Motorcycling and Motoring*, 9 April 1902, p. 147; *Motorcycling and Motoring*, 7 May 1902, p. 216, two-cartoon sequence; See also *Motorcycling and Motoring*, 4 February 1903, p. 521, a copy of a cartoon appearing in a German magazine showing armoured cars, with bayonets etc.
[52] *The Car Illustrated*, 23 August 1911, cover.
[53] The board included the Hon. John Scott-Montagu (ten shares); the Marquis of Granby (Peer, twenty-five shares); Jarrott (Director, one share); Levitt (Spinster, one share); and Edge (Director, one share). See 'Agricultural petrol motor', *The Motor*, 23 February 1904; *Motorcycling and Motoring*, 3 September 1902, p. 62. See also *The Motor*, 29 April 1903, p. 323; and *The Autocar*, 24 October 1903, p. 506.

engine to drive machinery;[54] farmers and rural-based dwellers in the United States had already shown much imagination in using their cars such as the Ford Model T, for driving machinery by day and for pleasure touring during the weekend.[55] In 1917, during the First World War, Edge was appointed as Director of Agricultural Machinery in the Ministry of Munitions. In this capacity, and in a vain attempt to keep the foreign product out, he arranged for a rigged competition to 'demonstrate' the superiority of domestic tractors – there were many brands of tractor to choose from – versus American Fords. (Yet, it was 1950 before the number of tractors exceeded farm horses.)[56]

The potential at election time for budding politicians being able to address several gatherings of far-flung electors in one evening was not lost on some MPs, especially those in rural areas; R. Leicester Harmsworth, Liberal MP for Caithness since 1900, and brother of Alfred, was one of the first to use his car for electioneering.[57] (*Punch* saw the matter rather more cynically. Its cartoon, coinciding with the January 1906 election, showed an electioneering motor car having run somebody down. The passenger looked back and said, 'It's all right. Drive on! He's voted!')[58] Meanwhile, Frank Banfield observed in 1903 that 'the motor car proved itself a valuable political factor in Ireland last year', enabling the Earl and Countess of Dudley to see more of the 'wild rugged country of Connaught and its people'.[59] Prime Minister Arthur Balfour, a keen motorist, 'has expressed strongly his sense of the part automobilism may play in the elucidation of many of our social problems'. He had in mind, for example, the clearing of 'unsatisfactory areas' in towns, to prohibit overcrowding 'and to make new thoroughfares and to widen old ones'. 'Unless I am greatly mistaken, the most effectual means for dealing with congested populations and congested traffic will be found in the cheapening and improvement of methods of locomotion.'[60] Montagu also declared in 1903:

> Those who have studied the subject [motor traction] and take the trouble to look a little ahead realise that the motor-car will be in a short time of immense service to the great mass of the people; that it will prove an important factor in the clearing of congested areas; that it will reinforce dwindling rural populations; that it will bring prosperity to the small agriculturist, supplant or supplement inadequate railway services, besides conferring health, pleasure and increased facilities of transport on thousands of our fellow countrymen.[61]

[54] *The Car Illustrated*, 15 April 1914, p. 346.
[55] See James J. Flink, *America Adopts the Automobile, 1895–1910* (Cambridge, MA: MIT Press, 1970); and more recently, John Heitmann, *The Automobile and American Life* (Jefferson, NC: McFarland, 2009). For a sense of how the Ford Model T was adapted in the United Kingdom, see Chris Barker, Neil Tuckett and Drew Lilleker, *The English Model T Ford, Vol. 2: Beyond the Factory* (Keighley: Model T Ford Register of Great Britain, 2014).
[56] Thompson, 'Nineteenth-century horse sense', p. 63.
[57] *Motoring Illustrated*, 12 April 1902, p. 131.
[58] *Punch*, 17 January 1906, p. 46.
[59] Frank Banfield, 'Who's who in the motor world', *Harmsworth London Magazine*, 10 (1903), pp. 525–36, at p. 527.
[60] Banfield, 'Who's who in the motor world', p. 526.
[61] 'Try to understand please', *The Car Illustrated*, 20 May 1903, p. 387.

The light car for 100 pounds

Such mobilized futures could only develop with an engaged and much wider user base. Magazine editors (particularly Dangerfield for *The Motor*) and entrepreneurs were keen to emphasize the viability of the cheaper end of motoring although the offerings were often rather nasty and undeveloped in the pre-First World War period. For these aspiring motorists, the full-size motor car for 100 pounds had long been on the horizon, breaching a psychological price barrier, and the magazines believed it would represent a sales breakthrough. In 1903 Edge addressed the Royal Norfolk Automobile Club with his paper 'Motors, past, present and future'. He had noticed that the public wanted a car costing about 100 pounds, 'light, cheap to run, cost little for upkeep, capable of carrying from four to six persons, with room for luggage, a hood to keep off the dust and rain, and which could be converted into a brougham in very cold weather, or if a theatre or dance was attended'.[62] He was tapping into the aspirational world conjured up by *The Motor* and the other motoring weeklies.

It was already possible to purchase a car for about 100 pounds: it has been seen in a previous chapter how in 1902 the American Crestmobile, imported by the O'Halloran Bros, was available for 100 guineas (105 pounds), even with payment by instalments. In its coverage of the Paris motor show, *The Motor* drew up a report on all light cars, domestic and foreign, then available. Of the thirty-odd British varieties, the A. C. L. Voiturette was priced at 105 pounds, while the New Orleans 'standard 3hp' was 100 pounds (or 105 pounds with a nine-inch longer wheelbase, spare petrol tank and touring basket). There was also a new model, the Bijou (price yet to be fixed, but 'will be considerably under 100 pounds'), the Hind ('well under 100 pounds'); and the Endurance Voiturette for 100 pounds, fitted with pneumatic tyres.[63]

Yet, the 100-pound light car for the early period remained elusive – the Hind and the Endurance disappeared, while the Bijou survived only a year.[64] The Miniature Velox, introduced in 1902 and costing 125 pounds, suffered a similar fate, lingering until 1904, with only twenty-one made.[65] There was Humber's light car of 1903, at 120 guineas (126 pounds)[66] while Vauxhall's new light car for 130 pounds (plus the front seat) was 'an excellent vehicle for the man of moderate means'.[67] However, the 1903 6hp De Dion Voiturette at 200 pounds, imported by (Edge's) De Dion Bouton British and Colonial Syndicate Ltd, was probably the benchmark, both for its quality and the minimum price a customer should pay.[68]

---

[62] *Daily Mail*, 12 December 1903. Edge showed an interest in developing small cars, for example, Cubitt and AC, but not until after the First World War.
[63] *The Motor*, 3 December 1902, pp. 295–309, and cont. 10 December 1902, pp. 323–5; *The Motor*, 17 December 1902, p. 349.
[64] Georgano, *The Beaulieu Encyclopedia of the Automobile*, i, p. 165.
[65] Featured in *The Motor*, 3 December 1902, pp. 298–9; and *The Autocar*, 27 June 1903, p. 756. See Georgano, *The Beaulieu Encyclopedia of the Automobile*, iii, pp. 1671–2.
[66] *The Motor*, 10 June 1903, pp. 401–2.
[67] *The Motor*, 12 August 1903, p. 14; *The Motor*, ads, 26 August 1903.
[68] 'The new popular De Dion Voiturette' is described in *The Motor*, 14 January 1903, p. 426–7. Edge, Jarrott and Duncan formed the De Dion Bouton British and Colonial Syndicate Ltd in 1899 with a capital of 10,000 pounds. Edge was the managing director, Jarrott the manager and Duncan the

The gap between expectation, price and reality, then, persisted. *The Motor*, in its editorial 'Cheap cars', found that 'Ideas as to what is a cheap car must vary with the outlook, or perhaps we should say, the income of the individual'. It concluded that

> the greatest potential demand unquestionably is for a machine with two seats side by side at about 110 pounds to 125 pounds, and there is a very large section of the public which will regard anything above this price as being otherwise than cheap, whatever the makers may choose to call it ... nothing would please us better than to see a good selection of such machines available.

Those waiting must be patient because they either do not exist or 'the examples are of inadequate construction'.[69]

Advertising played a key part in sending a message that something which was neither motor-cycle nor motor car should be thought of by the intending consumer as a fully fledged motor car. The 'forecarriage', otherwise forecar or 'trimo', was described in an article by B. H. Davies in 1903 as a 'new arrival' with much promise. Essentially a motor-cycle whose front wheel had been replaced by a two-wheeled bath chair to carry one passenger ahead of the driver, Davies described variations on this theme (and also related the story of one front attachment coming adrift at 18mph). It concluded that the motor-cycle and forecarriage 'constitute a most simple and fascinating vehicle for two persons.' A 2¾hp engine gave 20mph, the front passenger got no exhaust or dust, conversation would be 'easy' and comfort 'considerable'. It advised, do not expect it to trot two passengers up a 'danger' hill, so drop off the passenger at the bottom (making him or her walk up), 'sail up yourself, and wait at the top, thus cooling the cylinder and enjoying a cigarette'.[70] So, if the light car was barely available for £100, entrepreneurs saw opportunities in elevating the lesser machinery – motor-cycles and 'trimos' – then available. The Coventry Eagle 'Trimo' was billed as 'Almost a motorcar!'[71] Often with engines of about 3½hp, forecars were the natural mid-point between motor-cycling and light cars; the Phoenix Trimo, marketed by cycle- and motor-cycle manufacturer Joseph van Hooydonk (1866–1936), was sixty pounds.[72] Similarly, there was the Eagle forecar, 'The most successful two-seated car ever built.... Starts like a car, goes like a car, it is a car, but – it is half the price of a car'.[73] Other alternatives cropped up, such as the 3hp twin-cylinder tandem motor-cycle for £75.[74] These developments – a move towards the light car for the many – did not go unnoticed outside the motoring-magazine bubble, with, for

---

manager of the Paris office. Other cycling entrepreneurs associated with the company included Stocks, Munn and Hubert Egerton: Duncan, *The World On Wheels*, ii, p. 812.

[69] *The Autocar*, 2 May 1903, p. 509.
[70] B. H. Davies, 'The forecarriage and how to drive it', *The Motor*, 3 June 1903, pp. 369–70. A 'danger' hill is one that warranted a sign being erected to warn cyclists and motorists.
[71] *The Motor*, 8 July 1903, ads. 'Trimo' became a generic term for forecars, while 'quad' became a term for four-wheeled forecars.
[72] *The Motor*, 8 July 1903, ads.
[73] *The Motor*, 8 July 1903, inside front cover; and 15 July 1903, ads: 'It is a car, not a hybrid motor-cycle'. Eagle used 4½, 6 and 9hp De Dion engines.
[74] *The Motor*, 8 July 1903, ads.

Figure 7.2 'Almost a motorcar!' The 'trimo' was a generic name for a motor-tricycle with the passenger at the front, and which was popular in about 1903. This advertisement promised motor-car luxury at the price of a motor-cycle: *The Motor*, 8 July 1903, ads, n.p.

example, the *Illustrated London News* picking up on the possibilities and publishing an article, 'Motors for the Million' in 1905.[75]

The world of the sub-£100 motor car was, then, an altogether different one to that described by Harmsworth in his Badminton book, or of art-paper magazines such

---

[75] Jeremiah, *Representations of British Motoring*, pp. 19–20.

as *The Car Illustrated*, or *Motoring Illustrated*. It was a different world to that of the man whose application to join the Automobile Club would be acceptable. Of course, fashions proved to be fickle. *The Motor* in an editorial soon recognized that the forecar was flawed, with its one-wheel drive. It speculated elsewhere in the same issue that motor-cycling for two was ideal, so wondered whether we could expect in time to see the tandem light car.[76] (*The Motor*, then, predicted the cyclecar boom, described later.) The gap between the forecar and the light car remained to be filled. 'Expectant' wrote in 1905 to say, 'I represent a very class of people who have no room for a car, and could not afford either its primary cost or its upkeep, yet desire something better than either the forecar or the sidecar'. He thought of a tricar, with a 5hp, air-cooled two-cylinder engine, weighing 3cwt and costing £90–100.[77] The commentator Henry Sturmey, in his 1910 article, 'That £100 Car', found it still eluded us, but he reported on new efforts such as one by a Mr McMullen, with a four-cylinder two-seater for £120, and which had had a good independent report.[78] In 1910 'Runabout' in his regular feature 'Small car talk' in *The Autocar* wrote how he occasionally received 'pathetic' letters from motorists who could not afford the small Rovers or Phoenixes and wondered why they could not get something for £100 ready for the road. It has been attempted in the past, 'Runabout' noted, and presumably had not been successful because men want more for their money than the £100 car will give them and are not content with the 'modest specifications'.[79]

'New motoring'

The 'breakthrough' came in about 1910, with what the press called the 'new motoring' boom. It was clear a new type of vehicle was evolving, a very simple and crude single- or two-seater motor car, and formally named 'cyclecars' by the Auto-Cycle Club (ACC).[80] One immediate difference in the promotion of the cyclecar was the invisibility of the entrepreneur in the process; established 'names' were not getting involved as agents. Often it was the same company doing the assembling and the marketing, for example, the Portland or the Kendall, while the address given for the French Automobilette was that of its makers in Paris: for this cyclecar, then, there was no agent or importer.[81] Technically, cyclecars were as diverse as the first light cars had been ten years before:

[76] *The Motor*, 15 March 1904, pp. 145, 153.
[77] *The Motor*, 10 January 1905, p. 636.
[78] Henry Sturmey, 'My way of thinking', *The Motor*, 1 March 1910, p. 153. This car was subsequently marketed as the Rational: *The Motor*, 24 May 1910, p. 634.
[79] *The Autocar*, 22 January 1910, p. 119.
[80] The ACC identified a cyclecar as one of under 6cwt or 1100cc, an arbitrary limit for competition purposes: *The Motor*, 19 October 1912, p. 681. For cyclecars, see C. F. Caunter, *The Light Car: A Technical History* (London: HMSO/The Science Museum, 1970), esp. chap. 4.
[81] A sample taken from advertisements of *The Cyclecar* show, for example: the Bédélia's 'sole concessionaire' was a garage in Tooting; the L. M. gave as its address the L. M. Works in Clitheroe; the Portland's 'sole concessionaires' were the Portland Motor Co. of Great Portland St, London; the Kendall Carette was from Kendall Motors of Birmingham; and the Automobilette gave as its address its makers A. Coignet and J. Ducruzel, Constructors of Billancourt, Paris: *The Cyclecar*, 4 December 1912, ads supplement, pp. i–v.

Figure 7.3 The cyclecar ('new motoring') was a phenomenon of the period immediately before the First World War, and provided very cheap and cheerful motoring. This cartoon shows prevailing attitudes to women who were seen as potential 'new motorists'. Image: *The Cyclecar*, 27 November 1912, p. 21.

three or four wheels; drive by chain, shaft, friction or belt; and seating abreast or in tandem. The breed was typified by the Bédélia (a four-wheeler, front-engined device with tandem seating, and the driver at the rear), which 'claim[ed] to have solved the £100 car problem',[82] or the GN (four-wheeler, front engine, two-abreast seating), with dozens of hopeful makes springing up in the period from about 1910 to the First World War. The immaturity of the product was apparent by the need for magazine articles such as 'Three wheels or four?', and by letters asking if a differential was necessary or even desirable; whether to opt for belt or chain drive; and with readers offering their own home-grown variants.[83]

By about 1910 the cover illustrations of *The Motor* had for a few years been depicting scenes to suggest a move upmarket for the magazine, away from the 'modest motorist', to join the comfortably off middle classes in their detached properties on tree-lined streets. The cyclecar boom, then, presented the editor Dangerfield with another opportunity, to introduce another spin-off magazine, *The Cyclecar*, and here Dangerfield had placed himself ahead of the competition, as it was another twelve months before Iliffe, publishers of *The Autocar*, responded, with their monthly *The Light Car*, at 3d and with a clear '*Autocar*' look. With its style clearly mimicking *The Motor*, 80,000 copies, and then a reprint of 20,000 of the first edition of *The Cyclecar*

---

[82] *The Motor*, 15 March 1910, p. 242; *The Motor*, 29 March 1910, p. 318, on sale for £48. Cyclecars included typically, the Portland (£65); AC Sociable (£92 10s); Sabella (£99); Premier (£100); and the GWK (£135). The new 'Bullnose' Morris Oxford was £175 (all from *The Cyclecar*, 25 December 1912, ads and inside back cover, and all were basic prices, usually excluding windscreens, hoods etc.).

[83] See, for example, *The Cyclecar*, 25 December 1912, pp. 149–54.

were printed, edited by Dangerfield and costing a penny.[84] Dangerfield wrote on its launch that, even though there were at that point probably no more than one hundred owners of cyclecar-type vehicles out there, the cyclecar must 'of necessity' make a strong appeal to an enormous new class of people, who had never owned either motor-bicycles or motor cars, the former because of their restricted comfort, the latter because of their price.[85] He commissioned an article by the Liberal MP L. G. Chiozza Money to calculate, from the official returns of the annual rentals of private dwellings, how many people in Britain could afford a cyclecar. His assumptions were rather hopeful, and assumed the possibility of more than one cyclecar user to a household, such as the inhabited upper parts of a shop, those who live in chambers, flats, hotels and so on. Taking houses with an annual rental of £41–100 gave 336,000 households, the heads of which earned £400–1,000. Some could also be expected to be drawn in from the £21–40 house-rental bracket (1,088,000), and excluding some of the top end who probably already owned (or could own) a 'motorcar proper', then 'at least' 350,000 could afford a cyclecar.[86]

The cyclecar hype was electrifying. Edward Shrapnell-Smith suggested the cyclecar would be useful as an 'auxiliary', a handy second vehicle for minor uses where the big car was not required. In this attempt to brand the cyclecar as a 'classless' car, the 'wealthy automobilist' would have many cars, used for different purposes. One suggestion was that the highlands, with its terrible roads, would suddenly be opened up by a narrow, handy car such as the cyclecar; and here, the doctor need no longer leave his vehicle on the high road to trudge up the path, to avoid wear and tear to his car. In short, the 'bugbear of cost has disappeared: the cyclecar, for its total running and maintenance, will often cost less than only the pneumatic tyres of a larger and more powerful motorcar.'[87]

The illustrations appearing in *The Cyclecar* and on its covers emphasized this image. One drawing showed a cyclecar being driven through (infeasibly wide) gates onto the path of a front garden.[88] Another showed one with five adults aboard,[89] while another had an entire family fitting in.[90] 'A Cyclist of 18 Years' wrote about its suitability as a single-seater. He had never liked motor-cycles ('weight, clumsiness and complication') and always cycled solo ('having no friends or relatives whom I wish to take out'), so the single-seater cyclecar was ideal.[91] To emphasize the cyclecar's potential, a doctor was pictured paying a call in one,[92] while it was also apparently ideal for going camping,

---

[84] *The Cyclecar*, 27 November 1912, p. 13. It became *The Light Car and Cyclecar* within twelve months, that is, like *The Motor*, the start-up title lasted barely a year. For a broad survey of cyclecars, see Michael Worthington-Williams, *From Cyclecar to Microcar* (London: Dalton Watson, 1981). See Armstrong, *Bouverie Street to Bowling Green Lane*, pp. 104–10.

[85] Armstrong, *Bouverie Street to Bowling Green Lane*, p. 108; *The Motor*, 22 October 1912, p. 476.

[86] L. G. Chiozza Money, 'Welcome to a new industry! Gigantic possibilities for the cyclecar trade', *The Cyclecar*, 27 November 1912, pp. 11–12.

[87] 'E. S.-S.' [Edward Shrapnell-Smith], 'The cyclecar as an auxiliary: a handy vehicle for minor uses where the big car is not required', *The Cyclecar*, 27 November 1912, pp. 20–1.

[88] *The Cyclecar*, 27 November 1912, p. 4.

[89] *The Cyclecar*, 30 April 1913, cover.

[90] *The Cyclecar*, 22 January 1913, cover.

[91] *The Cyclecar*, 18 December 1912, p. 94.

[92] *The Cyclecar*, 29 January 1913, p. 275.

with the carefree owner pictured relaxing by playing the banjo.[93] It was useful for golf (motor cars were unnecessarily big, while motor-cycles made it difficult to carry the 'sticks'); angling (to get to spots remote from trains, where heavy motor cars were no good); as well as boating, shooting and getting to and from the cricket, football or tennis fields.[94]

There was a reason why the cyclecar was cheap and lightweight, though, and the reality was that most were 'downright evil to drive even when new'.[95] The Cyclecar Club wrote a damning report following the Reliability Trial held in 1913, in which many failed the hill-climb, and were deficient in fittings, brakes, driving, luggage-carrying, noise and gears.[96] The English Six Days' Trial later that summer, in the Lake District, on poor roads and with rain on one day, was reported as a 'Farcical test to destruction' where of its twelve entrants only three finished (according to *The Cyclecar* this was, apparently, all the fault of the organizers, the ACC, who had organized a 'wholly unsuitable course' and as a motor-cycling body could not have been interested in cyclecars).[97] Letters by concerned readers, and by those manufacturers whose entrants finished, challenged the magazine's stance, asking what the point was of a contest if everyone passed, and pointing out that most of the withdrawals were due to engine failure.[98]

Caveat emptor, then. But a decade had passed since the 'modest motorist', and now an altogether more savvy consumer was watching the market. An article on the costs of a cyclecar highlighted how 'for some the stumbling block is affording the initial outlay, for others it's the running costs'. For the former,

> the remedy is in sight. Several firms of good standing are prepared for quite a modest initial outlay to supply the cyclecar, providing an agreement be entered into for the payment of the balance of the purchase money at the convenience of the buyer. The writer holds no brief for the credit and instalment system. It would be better far if it could be avoided.

So, payment on 'terms' was available to purchase a cyclecar, just as it had been for the earlier light cars for 'modest motorists'. As for minimizing running costs, instead of using the cyclecar for a long weekend (Friday night to Monday morning), the man 'of limited means' might have to be content with a Saturday afternoon run and a longer one on Sunday, or maybe just the Sunday run. The definition of a run may have been as little as thirty or forty miles there and back – reflecting a more realistic situation for most users – or it could be 250–300 miles with hotel accommodation and garaging. The cost per mile was generally 2d.[99]

---

[93] *The Cyclecar*, 28 May 1913, p. 5.
[94] 'The cyclecar and other pastimes. A brief indication of its utility', *The Cyclecar*, 1 January 1913, pp. 155–6.
[95] Reg Winstone, 'Grand Prix cyclecar', *The Automobile*, August 2019, pp. 22–30, at p. 30.
[96] *The Cyclecar*, 26 March 1913, pp. 469–71.
[97] *The Cyclecar*, 27 August 1913, p. 354.
[98] *The Cyclecar*, 27 August 1913, pp. 345–53.
[99] *The Cyclecar*, cf. pictures on covers, 7 May 1913, pp. 617–8.

The Motor, optimistic as ever, believed there was now something for all pockets. If the purchaser could not go above £100, he would go for a 'type of machine' which was in truth a cyclecar or an 'edition de luxe' of the motor-cycle and fixed sidecar, the difference being that he had a comfortable protected seat for him and his companion, and probably four wheels instead of three. If he could afford £150–180, he would be guided by whether he wanted one companion or three. If he opted for a two-seater, he would probably buy a 'miniature car' which may well come into the ACC definition of cyclecar, but if he wanted four seats he would have to give up a cyclecar altogether 'and go in for one of the wonderfully cheap productions which are now on the market'. There would be no 'sharp line of demarcation':

> The users of cyclecars, runabouts, miniature cars, small cars, light cars, call them what you will, will be no new class, but will be mainly drawn from the ranks of motor cyclists and motorists. We do not say that cyclecars and miniature cars will attract no new recruits to motoring, because they will, for the simple reason that the type, in one or other of its varieties, will fill a price gap which has existed between the passenger motor-cycle and the small motor car. On the other hand, the owner of a miniature car will be just as much a motorist, and often more of a motorist, than the owner of a so-called lordly limousine. There is not going to be, as some people imagine, a new class of road user who is neither a motorist nor a motor cyclist, but a 'cyclecarist'. How can there be, when, as we have shown, at one end of the scale, the machine will be hard to distinguish from a two-seated motor-cycle, while at the other no one will be able to say definitely when the cycle car becomes a motor car.[100]

The reality was that the cyclecar filled a gap and created more motorists in the few years before cars such as the Ford Model T (Manchester-built from 1911) and the 'Bullnose' Morris Oxford (1913) appeared, both of which looked more like large and conventional cars, but shrunk down, and had immediate appeal. But their prices never matched those of cyclecars (the Bullnose was £175, the Model T from £125).[101] However, cyclecars – the 'new motoring' boom – were a fleeting phenomenon, and showed that quality, reliability, standardization and the £100 light car remained mutually exclusive.

Imagining the road network

Before cyclists started to venture onto the road network from the 1870s, horse-based traffic and pedestrians used the roads at walking pace. Such users had no need for warning signs, the 'correcting' of adverse camber, or the straightening out of 'blind' bends. The pace and nature of traffic before the bicycle and motor car meant that little dust was kicked up (and would otherwise have been laid by the water cart). Meanwhile,

---

[100] *The Motor*, 19 October 1912, p. 681.
[101] *The Light Car and Cyclecar*, 5 January 1914, ads, p. 17 (Morris); *The Autocar*, 22 November 1913, back page (Ford: American 2-seater).

within towns, traffic congestion had long been a fact of life.[102] But cyclists – and motor cars once they started appearing in numbers from the 1890s – found the road network unsuited to their new purpose. Agitation for road improvement had started with the cyclist pressure-groups of the 1880s; Reid has shown how cycling groups had long been lobbying for road improvements, leading to the Roads Improvement Association in 1889 with W. Rees Jeffreys (1871–1954) as its secretary, who later served as secretary of the Automobile Club.[103] An increasingly vocal and powerful motoring lobby highlighted new 'problems', such as the condition of roads, the absence of signposting and the presence of 'dangerous' hills. Such lobbying played its part in the establishment of the Road Board by the then chancellor of the exchequer David Lloyd George in 1909. Different road surfaces were tried,[104] and while some roads were 'improved', congestion was soon a new feature, particularly of popular routes such as the London to Brighton road. A reader of *The Autocar* counted passing traffic through Haywards Heath (between London and Brighton) in 1910 at several times on weekdays and weekends, and conclusively showed that something like double the number of vehicles used the road at weekends; about 200 vehicles passed through on a Sunday, the busiest day.[105] In 1913 Henry Sturmey, saying he lived on 'one of the lesser main roads' of the Midlands, had observed that ten years before a horsed vehicle was a 'rarity', with maybe five or six an hour at the most, but that now, rarely more than a few minutes elapsed without a car passing.[106] It did not take much new motor traffic, then, to clog up the popular routes.

With roads like these in mind, Claude Johnson wrote in 1906 that the 'watchword' for the motoring pioneers, once 'the road and its discovery', now had to be shifted to 'the road and its improvement'.[107] Dedicated motor roads had been anticipated from the earliest days, sometimes with uncanny accuracy as in H. G. Wells's *Anticipations* (1902).[108] Motor roads, especially to the south coast, were most anticipated; Montagu believed that a London to Brighton highway had more commercial promise than going north, but his efforts in parliament were strangled by the rail lobby in 1906. Motoring-magazine editors commissioned artists' impressions of new roads for motoring, often fast, straight, multilaned, with horse, pedestrian and bicycle traffic disallowed. A drawing of another 'motor way', with three lines of traffic in one direction, including a

---

[102] See Morrison and Minnis, *Carscapes*, chap. 12, esp. pp. 319–23.
[103] Reid, *Roads Were Not Built for Cars*. For Jeffreys, see E. K. S. Rae, 'Unconventional portrait of leaders in motorism: Mr W. Rees Jeffreys', *Automotor Journal*, 10 July 1909, pp. 824–7.
[104] Reid, *Roads Were Not Built for Cars*, chap. 7.
[105] Arising from a recent court case, G. T. Langridge wrote to Revd H. R. White regarding the volume of traffic passing through Haywards Heath in 1909. White had claimed in court that 500 cars passed through per weekday and 1,000 on Sundays. Langridge conducted his own survey of the Crawley Road, revealing:

> Wednesdays: 26 May 1909: 104 cars; 28 July 1909: 92; 22 September 1909: 79
> Sundays: 30 May 1909 (Whitsun): 208; 1 August 1909: 139; 26 September 1909: 211

> Taking in return journeys, Langridge concluded this gave a weekday average of 184 and, for Sunday, 372: *The Autocar*, 26 February 1910, p. 285.

[106] Sturmey, 'The amazing growth of the motoring movement', p. 240.
[107] Claude Johnson in Montagu, *The First Ten Years of Automobilism*, p. 29.
[108] Morrison and Minnis, *Carscapes*, pp. 239–40; this is the one illustrated in Jeremiah, *Representations of British Motoring*, p. 240.

motor-cyclist and a car with streamlined, and therefore futuristic, bodywork, appeared in *The Motor* in 1905.[109] With the proposal for a new road from London to Brighton in 1905, and with spurs off for Folkestone and Portsmouth, *The Car Illustrated* presented an artist's impression. Here, a new, multi-carriageway road flew over a minor road, on which walkers, cyclists and horse-and-cart users ambled.[110] The same magazine, in 1914, identified a 'need' for an outer-circle road for London and published a map with a projected London circular remarkably similar to what would be the M25, completed in the 1980s.[111]

However, while imagery of a future of fast, dedicated highways was evident, such roads did not necessarily have wide appeal, even amongst motorists. In the article, 'Special roads for motorists: Are they wanted?', *The Motor* mused that they would serve as a 'weapon' for 'our enemies'. The 'motor movement has already been held in check by the class which everlastingly bristles against innovation and opposes progress'. And think of the 'dreary monotony' of driving on these roads.[112] Letters to motoring magazines suggested a similar train of thought. Eric W. Walford was unimpressed with the suggestion of a 'special' road from London to Brighton as it would show that motor cars were not safe on the ordinary road. Furthermore, such roads would be used as a 'scorching track'.[113]

Some infrastructure improvements were made, mostly to straighten roads, but roads remained unfit for the purpose. Yet, Minnis has shown by the use of photographs of English roads taken from the air in the 1920s that there remained few visible signs of change to the roads' infrastructure, or indeed, of the incursion of the motor vehicle.[114]

The persistence of the horse

With the arrival of the railway in the 1830s, the 'downfall of the horse and the horse interest' had been predicted, and while horse numbers did momentarily fall, the railways, instead, *expanded* demand for horse labour. By 1890, there were 6,000 railway-owned horses, all requiring elaborate stabling. Indeed, as late as 1934, some 14,834 horses were owned by the four main rail companies (and another 12,605 by the army).[115] The same downfall was anticipated in the 1890s with the arrival of the motor car. The 1896 'Emancipation Run' from London to Brighton had, in part, been predicated on the basis that it would demonstrate how the motor car would render the horse useless thereafter, and there was widespread relief in the newspapers when it so clearly failed to demonstrate anything of the kind. Yet, the possibility clearly remained

---

[109] *The Motor*, 21 November 1905, p. 431.
[110] 'Proposed motor track carried over an existing road', *The Car Illustrated*, 25 October 1905 (this is reproduced on the dust jacket of Jeremiah, *Representations of British Motoring*).
[111] *The Car Illustrated*, 29 April 1914, pp. 406–7.
[112] *The Motor*, 14 November 1905, p. 371.
[113] *The Motor*, 21 November 1905, p. 450.
[114] John Minnis, *England's Motoring Heritage from the Air* (London: Historic England Publishing, 2014), *passim*.
[115] Thompson, 'Nineteenth-century horse sense', pp. 64–6.

in the popular imagination over the next decade. As late as 1905, a *Punch* cartoon showed a 'Stinkum' motor lorry ferrying a horse on a flat platform, with a sign on the side, 'The passing of the horse: No further use for him.' In the following picture, the lorry had broken down and was being towed by the smug-looking horse.[116]

The horse and its abilities were the standard by which the motor car was judged. Motoring entrepreneurs, to promote their novel product, worked hard to demonstrate how the motor car could 'do' everything a horse could do, and more. Stunts, carefully planned and rehearsed, showed a motor car, for example, negotiating the steps outside a town hall.[117] E. H. Coles was the 'expert driver' at Benz agents Hewetsons in 1900 and it was advertised that he would do 'fancy autocar driving' at the Alhambra in an 'Ideal' Benz. This involved ascending a one-in-four incline to a small platform, then down a one-in-three: 'If anything is calculated to convince the general public of the handiness of motor cars, surely this is it.' Jeremiah has shown how the Daimler Motor Co. used the successful ascent of Worcestershire Beacon in 1897 (425m) as a clear promotional technique in its sales literature. In addition, in 1911 Edinburgh Ford dealer Henry Alexander promoted the Model T by driving one to the summit of Ben Nevis (1,345 metres);[118] of course, staging these stunts required careful preparation, a clearing of all obstacles in advance, and a priming of the press. Being able to ascend hills was a prerequisite for sales of motor cars at the time. It was a frequent subject for letter writers to motoring magazines, while local clubs put on hill-climbs and other trials (and key sections of the Thousand Mile Trial of 1900 deliberately included the ascent of long hills). Agents made much out of this ability – a glance through the ads section of any motoring magazine will attest to this.[119] And if the motor car could ascend hills, it would be fully able to deliver the owner to his garden gate, or the door of his office.[120]

So for those same entrepreneurs, a future needed to be peddled where the all-capable motor car would supplant the horse. Edge, guest-editing the *Penny Illustrated Paper* in 1913, mused, 'Some people say that as regards the motor car, it has nearly reached its limit of business development. But this is certainly not the case. So long as

---

[116] *Punch*, 17 May 1905. See *The Autocar*, 25 February 1905, p. 305, for the photo on which this cartoon was based: a Lacre van with a horse in the back, and the sign 'Unemployed, the passing of the horse: Lacre Industrial cars supersede horse-drawn vehicles'. Lacre 'sent out at intervals' the van to drum up interest during the Olympia show, and 'the passage of so unique a turnout through the streets of London naturally attracted a great deal of attention'. (It is unclear whether the Lacre broke down.)

[117] See E. Doan, 'Some wonders from the west: LVI: Riding upstairs in an automobile', *The Strand*, 25 (1903), pp. 593–4, for an account of the publicity stunt of driving an Oldsmobile up the twenty-five steps (44 per cent incline) of the State Capitol building, Lansing, Michigan. This tale was evidently well known in the United Kingdom because the Oldsmobile agents advertised one under the heading, 'Will you stand still or move forward?', with a drawing of an Oldsmobile ascending steps that disappear into the distance: *Motorcycling and Motoring*, 19 November 1902, ads. See also Duncan, *The World On Wheels*, ii, pp. 890–2, featuring a sales initiative by Charles Jarrott and Letts, importers of the Oldsmobile, in which the car performs the same stunt.

[118] *The Times*, 18 May 1911. For footage see www.youtube.com/watch?v=jaNgYhvmtzA, accessed 30 July 2019. My thanks to John Harrison for these references. See also Chris Barker, 'The rise and fall of the Model T in Britain', *Aspects of Motoring History*, 12 (2016), pp. 35.

[119] For example, the entirety of the text for the International Motor Co. is 'Portland cars are marvellous hill-climbers', *Automotor and Horseless Vehicle Journal*, 2 May 1903, ads, p. v. On the following page, Cottereau cars are 'excellent hill-climbers' (p. vii), and so on. Napier used the phrase 'marvellous hill climbers' in its sales brochures for 1902 and 1903.

[120] Jeremiah, *Representations of British Motoring*, pp. 16–17.

you can see a horse in the street, the work of the motor car is not finished. So long as it is cheaper to walk or drive to your destination there is more for the motor car to do!'[121] People of influence made pronouncements about how the horse would now be less significant. When the Thousand Mile Trial passed through Manchester, an account of the lord mayor's speech reported how if the Trial had taken place just two years before,

> they would have been received with the jeers of the grown-up population, and probably the stones of the younger members of the community. There was a sort of notion then that the autocar was a wild unmanageable animal, with wild men seated atop, whose sole purpose it was to break legs, or to knock ladies down and tear their dresses.

The lord mayor's point was that he was not going to give up his horses, but was certainly going to take an interest in the new 'sport'. He mentioned how few horses had reacted adversely to the passing motor-car, unless the (horse-)driver was nervous. 'Certainly, horses were getting perfectly accustomed to motors, and they would know that to a horse the autocar was the greatest blessing that had ever been put upon the road. He thought the public had come to realize that the movement had come to stay.'[122] Motor-vehicle advertising, too, had to appeal to a new generation of motorists with no horse-driving or cycling 'heritage', and one method was to concentrate on an imagined effortlessness with which the motor car could be driven. One example is an advert for the Adams-Hewitt of 1906, which had 'no old-fashioned levers to handle. PEDALS to PUSH, that's all', and the illustration featured a driver who was semi-recumbent in his efforts to appear relaxed.[123] While the Adams-Hewitt turned out to be a footnote in motoring history, the UK-assembled Ford Model T, introduced in 1911, had a crunch-free epicyclic gearbox requiring little skill or mechanical sympathy to operate and which undoubtedly helped its phenomenal sales.

Some commentators anticipated that even in the face of the new motor traction, the horse would persist. Salomons, for example, said, 'it will probably take another twenty years to drive horses off the road, except for pleasure driving, riding and farm purposes.'[124] The 'problem' with the horse – its ongoing costs whether in service or not, the issues of manure on the public highway, and public health – had long been recognized. Horse-drawn omnibus operators in the 1890s experimented with electric and steam versions – all unsuccessful – before finding success with the petrol version. They were motivated by a need to keep costs down; Thompson has pointed out how a town-horse would consume 1.4 tons of oats and corn, and 2.4 tons of hay, per year, costing ten shillings a week, all whether the horse was in service or not.[125] Motoring magazines and writers – those with a vested interest in the 'success' of the motor car

---

[121] *Penny Illustrated Paper*, 8 February 1913.
[122] *The Autocar*, 5 May 1900, pp. 420–38 at pp. 426–7. F. T. Bidlake was unimpressed, saying the lord mayor 'Dropped a few platitudes on mechanical traction and horses. Doesn't know the pleasures of motoring. Loves horses', *Bicycling News and Motor Car Chronicle*, 9 May 1900, p. 15.
[123] *The Autocar*, 21 April 1906, ad. My thanks to Malcolm Bobbitt for this reference.
[124] David Salomons in Montagu, *The First Ten Years of Automobilism*, p. 10.
[125] Thompson, 'Nineteenth-century horse sense', p. 78.

– published accounts of the (always cheaper) costs of running a motor car compared with a horse, and Salomons had said, 'the less [sic] horses we have, the less dust'.[126] There was also the need to stable and feed horses every day, the issue of their relatively short working life, and the logistical difficulties faced by local authorities in removing dead horses from the road.[127] *Motoring Illustrated* pointed out in 1914 how nobody in London or any large city could have failed to notice how house flies were diminishing year on year, and this was equated with horses being largely replaced by the car – 'thus the motor car has indirectly conferred a great benefit upon the community'; there were fewer sore throats, and better general health prevailed.[128]

That same observer would have seen the change not just in mode of traction but also in the rapid introduction of motor traction on public transport.[129] The rise of the motor bus was rapid. Each horse-cab required two horses to keep it on the road all day; many rail passengers would use horse traction, once having arrived at their station. There were some 25,000 horse-buses in use in Britain by 1890, which rose by as much as 50 to 100 per cent by 1902; Thompson calculated that since it took eleven horses to keep each bus in service, London, with 3,696 horse-buses in 1902, must have had 40,000 bus horses at its peak,[130] a huge and costly investment. The first motor bus was operated in Edinburgh in 1898, with others starting elsewhere too later that year.[131] A key factor in the lowering of running costs of motor buses was the falling contract price for solid rubber tyres; in 1905 this was now 2d per mile for a double-decker, giving 'the signal for London bus operation really to begin'.[132] Thompson has shown that in 1903 London had 3,623 horse-buses and only thirteen motor buses, but by 1913 this had reversed, with 142 horse-buses and 3,522 motor buses.[133] Away from the towns the picture was more mixed, with horse traction lingering in the more inaccessible parts, while in the Lake District, Scotland, Devon and Cornwall, horse-traction networks continued beyond the First World War.[134]

However, moving over to the motor car was not a one-way street.[135] The motor car remained prone to breakdown and this very unreliability was lampooned in the language of (horse-)driving. For example, in 1902 in *Punch*, the first time Mr and Mrs 'Spool' used their new motor car to drive to the hunt, 'the wretched thing bolted for fifty or sixty miles before it could be pulled up'.[136] Even by 1913, J. C. Jones was writing

---

[126] Salomons in Montagu, *The First Ten Years of Automobilism*, p. 10.
[127] See Thompson, 'Nineteenth-century horse sense'; McShane, *The Horse in the City*; Turvey, 'Horse traction in Victorian London', pp. 38–59.
[128] 'How the flies have disappeared', *The Car Illustrated*, 15 April 1914, p. 319.
[129] Walton has pointed out how little research has been done on the coach, bus, charabanc, and that a 'bricolage' method is necessary: Walton, 'The origins of the modern package tour?', pp. 145–63 at pp. 145–8.
[130] Thompson, 'Nineteenth-century horse sense', p. 65.
[131] John Hibbs, *The History of British Bus Services* (Newton Abbot: David and Charles, 1968), p. 42.
[132] Hibbs, *The History of British Bus Services*, p. 52.
[133] Thompson, 'Nineteenth-century horse sense', pp. 60–1.
[134] Hibbs, *The History of British Bus Services*, p. 36.
[135] *The Autocar*, 3 March 1900, p. 201. Coles was quite a draw: photos appeared in 'Notes on the Agricultural Show', *The Autocar*, 28 April 1900, p. 399, of Coles driving up and down the trick-driving slope.
[136] *Punch*, Almanac for 1902, n.p.

in to *The Cyclecar* to paint a miserable picture of motoring; in six months, 'I have never had a run without some involuntary stop or other'. (The editor hastily appended a comment, 'This is obviously an exceptional case . . .')[137] Some who had tried motoring returned to their old ways, while the reality was that the motor car remained unreliable despite what the magazines would have readers believe. Gelber has suggested that for some doctors, driving a motor car now required paying much more attention, when hitherto the horse had navigated routes by memory.[138] In 1905 'General Practitioner' wrote to *The Autocar* as a 'poor man' now considering returning to horses for his medical journeys because of the cost of tyres and punctures. His two friends were considering the same, and all three 'wish[ed] we had never taken to motor cars', despite each paying £400 for their motoring.[139]

Conclusions

Motor traction held a future of fantasies. For the many, this meant visions of flying cars, of chaos and traffic, but for the few – those who had grown used to having the motor car for their own leisure – they saw no reason why their fantasy should change. Neither was to get quite what they imagined. But their visions were fuelled by what they saw or read about (the speed records, the increasing reliability, a new normality as they surveyed the subtle and not-so-subtle changes on their high street) and by what they saw in stories and artists' impressions (the Paris-Madrid 'race of death', or the motor car, say, taking the place of the horse in the polo ring). Developments in flight in the first decade showed how rapidly horizons could change: in 1902 Santos-Dumont was giving interviews discussing his belief that the balloon would one day be the way to routinely cross the Atlantic, but within the decade Harmsworth's prize money for flying a powered craft from London to Manchester had already been claimed. And here a clear cross-pollination between the various new sports (motoring, motor-boating, aviation) is evident, with the same personnel as either adventurers, entrepreneurs or both.

There were worries about sustainable supplies of oil, and, consequently, the devising of technical solutions (using alcohol-derived fuel) in the expectation that dependency on the horse would naturally and quickly dwindle, and thus free up farmland for its new purpose. Yet, the horse remained persistent. The 'Emancipation Run' in 1896 was supposed to reveal the new motor traction as the convincing successor to the horse, but, instead, it proved a calamity and the blow did not come. Or at least it didn't in the way expected. There was no agenda to eliminate the horse (the horse for many remained one part of an increasingly broad set of mobility solutions), although few imagined horse numbers would actually increase in the period to the First World War as they did.

---

[137] *The Cyclecar*, 29 January 1913.
[138] Steven M. Gelber, *Horse Trading in the Age of Cars: Men in the Marketplace* (Baltimore, MD: Johns Hopkins University Press, 2008), p. 46.
[139] *The Autocar*, 17 June 1905, p. 813.

But imaginations about what motor traction could be used for could be rather dull – motor traction, many of the first pioneers thought, was about facilitating leisure for the already leisured, and any visions of motorizing, say, commerce, or the utilities, while not absent, were nothing like as apparent. It took visionaries such as David Salomons to set up commercial vehicle trials, but that in turn relied on risk-taking entrepreneur-editors judging the market was right and expanding their magazine portfolios into the new areas, just as the entrepreneur-businessman such as Edge continued to speculate on matters as diverse as motor-boating and powered aviation.

Motoring for the many saw quite a few false dawns in the early period: 'modest' motoring, then the 'new motoring'. They all brought a broadening of the motor-traction experience, but all perplexed the commentators who did their sums and calculated that many hundreds of thousands of motorists would immediately come into the fold if only the affordable and reliable motor car could be produced. The 'new motoring' movement, which brought about the cyclecar, was also backward-looking in that it reintroduced mechanical diversification (for example, taking up the drive by chain, shaft or friction) when the consumer really wanted comfort and standardization; the Stellite of 1913, for example, 'a proper car in miniature' typified the new breed which came to replace cyclecars. That, and others such as the 1913 Morris 'Bullnose' Oxford and the Manchester-built Ford Model T, introduced in 1911, are the ones that are remembered now, not cyclecars. The Ford in particular was no longer really a 'light car', coming into a higher horsepower-tax bracket, but was made in hitherto unprecedented numbers and backed up with a nationwide network of spares, and had an immediate wider appeal.[140] Importantly, the first entrepreneurs, and their customers – the 'modest motorists', from about 1902, then the cyclecar motorists, from about 1910 – had created an environment in which Morris, Ford *et al* could operate: the consumer, the dream, the infrastructure were by then broadly in place.

Motoring had become an experience for many more (buying second-hand, taking the plunge and trading up from their bicycle, or whatever), but it was to be much later in the century before it would become an experience for the many. Most driving would be by owner-driver, and in this new world, it was already anticipated that there would have to be dedicated roads on which the horse and pedestrian – and probably cyclist – would be unwelcome.

---

[140] For the Stellite, see Clausager, *Wolseley*, pp. 96–8. For the Model T, see Martin Riley, Bruce Lilleker and Neil Tuckett, *The English Model T Ford: A Century of the Model T in Britain* (Keighley: Model T Ford Register of Great Britain, 2008). Caunter did not see the Model T as a 'light car' and had little to say about it.

8

Conclusion

The 'old brigade', those late-nineteenth-century pioneer motorists with their cycling backgrounds and flair for sport and enterprise, would have been disappointed that motor traction did not turn out as they supposed it might. One of theirs, Henry Sturmey, wrote in *The Motor* in 1914 how in the early days, that is, about the late-1890s, motoring was, indeed, strenuous but it was a pleasure for those who adopted it. The problem was, he reckoned, that the 'idle rich' took it up in the early years of the twentieth century, when it had become reliable and competent chauffeurs could be had, and when there were landaulets and limousines and you didn't need a bath to be presentable. Then society adopted the motor car, 'rushing hither and thither, helter skelter', and those days of pleasure were gone.[1] Another, Henry Hewetson, agreed. He wrote in 1906, 'In the early days of motoring it was far more enjoyable than now. Wherever I went I was well received, and had invitations from absolute strangers to stop at their houses; and it was really not until the big cars came in that there was so much trouble with the horses and the general bitter feeling of the public against motor-cars arose.'[2]

Theirs was the world of the club, the amateur and the privileged. For them, from the latter part of the nineteenth century, horse-driving and then cycling had always been immensely uncomfortable, but offered the sport of the 'open road', theirs to discover, and free from the 'tyranny' of the train timetable. The new motor traction simply added a new and exciting dimension to their mobility options. Leonard Larkin, writing in *The Strand* in 1913, recalled his first ride in a motor car, a twelve-mile ride in a Benz in about 1896. In its ninety minutes, it broke down four times, but the driver had a bootful of tools and delighted in every breakdown. With each he got dirtier, greasier and more happy. Larkin said that this driver now had a modern car, which never broke down, and he had a 'face of settled gloom'.[3] The motoring writer 'Owen John' reflected on his early experiences: 'one had to be either a rich man, a fool, or in the business to travel the country by car. . . . They were unreliable, uneconomical playthings, and one used them at one's own risk and treated their disadvantages and uncertainties entirely as an expensive hobby.'[4] Trying to explain why then the motor car persisted, as the historian Gijs Mom asks us to do, had much to do with the entrepreneurial spirit of this

---
[1] 'Motoring for pleasure and amusement', *The Motor*, 10 March 1914, pp. 220–1.
[2] Henry Hewetson in [Lord] Montagu, *The First Ten Years of Automobilism*, p. 56.
[3] Leonard Larkin, 'Motor-Cars: Yesterday and today', *The Strand*, 46 (1913), pp. 636–41.
[4] 'Owen John', *The Autocar*, 16 April 1926, p. 28.

'old brigade'. But in fighting what must have seemed so often a lost cause, they were, it seems, victims of their own success.

There was no plan. Those first entrepreneurs were as much agents in the development of motor traction as their first customers were. Collectively, they did much of the on-road development that the manufacturer could never have afforded to do. This brought with it a tolerable reliability, and a standardization of sorts which fed in to the confidence the intending consumer needed. By this process a critical mass of light-car motorists ('modest motorists') was created between the period 1900 to around 1905, added to by cyclecar motorists ('new motorists') from about 1910. A shift away from a clubby culture, and the increasing availability of used motor-cycles and motor cars at much more affordable prices, drew even more people into the motor-traction movement. But before a 'consumer' could be created, before the emergence of the owner-driver who had little interest in the mechanicals or in getting his or her hands dirty attending to roadside calamities, there needed to be a sea-change in expectation of comfort. The account of T. B. Browne (1873–1965) in *The Autocar*, of 'two hundred miles through the snow', a 38-hour feat of endurance from London to North Wales in December 1899, tackling frozen snow up to six inches deep,[5] appealed to the 'pioneer' with its requisite grit and stoicism, but within ten years the reader of the same magazine would be expecting to read about comfort and the certainties of arriving. This rapid change is reflected in the shifting treatment of the motorist by reactionary middle-class periodicals such as *Punch*. Larkin pointed out this change, saying how it was 'only a few years' since 'a certain reverend gentleman' was convicted of lashing motorists across the face with a whip as he met them, to a point now (1913) when a majority of horse drivers allowed his car room to pass (although he could still see there was room for further adjustment, as it was 'far from a large majority').[6]

Motoring magazines, ever optimistic, calculated that there was a latent critical mass of motorists out there, just waiting for the psychologically significant half-decent 100-pound motor car. Once that appeared, people would spontaneously jump in and kickstart the motor-traction movement. However, there always had been brand-new motor cars for sale at 100 pounds or so, yet they were never bought in sufficient numbers. The motor car was not going to sell itself – the idea had to be sold, and also the consumers had to feed in to the process too. Even with purchasing options such as buying on instalments, or cooperating with friends or neighbours to make a joint purchase, few took the plunge. Furthermore, few home-grown manufacturers were able to tool up to supply in sufficient numbers, leaving the early market for the most part to the foreign manufacturers. The motoring entrepreneur S. F. Edge held agencies for cheaper cars such as De Dion, but he knew the profits were higher for more expensive brands such as Napier, if only more sales could be made. Even after the First World War, Edge's interest in promoting cheaper cars such as Cubitt came to little.

The impact of motor traction, contrary to the rhetoric and bluff of the motoring financier Harry Lawson, was slow and uneven. With expectations that the horse would be replaced forthwith, Lawson's 1896 'Emancipation Run', from London to Brighton,

---

[5] 'Tours and Runs', *The Autocar*, 13 January 1900, pp. 27–8.
[6] Larkin, 'Motor-Cars: Yesterday and today'.

could only be a disappointment, exposing the product as immature and irrelevant. For three or four years after that, motor traction – despite the lifting of on-the-road restrictions – languished, while the national club, the Automobile Club, tottered from one disappointing public event to another. There was, no doubt, a ripple of excitement for people as they saw their first motor car, and out in the sticks this could easily have been 1905 or later. But, once they had seen a few, as they probably had in London by 1900 at the time of the Automobile Club's Thousand Mile Trial, the reaction to its start and finish is indicative of subsequent indifference. Most of the onlookers as the cars lined up at Grosvenor Place for the start of the Trial were cyclists. There were a few 'non-participating motorists' but only a few pedestrians too.[7] For this event, over thirty cars returned three weeks later for the processional stroll into London, but as F. T. Bidlake reported: 'Scarcely any interest is shown on the concluding stages. The tour does not end with a climax, but with just a little feeling of relief. The Muster [of the returning motor cars] on Horse Guards Avenue is evidently not expected. No one knows of it, no crowd has assembled. All is quiet.'[8] During the Trial, while reaction by onlookers was generally positive (and this probably meant many workers and schoolchildren had been granted time off to be able to stand at the roadside), the reaction in, for example, Leeds, is telling:

> Run rather ignored by local bigwigs. Mayor too busy to take notice of it. Gets dropped out of history a hundred years hence when this trial ranks as starting great initiation of the public use of the then universal means of road travel. See Mayor in garish carriage, with vast parade of flummery.[9]

It was the one-time cycling racer F. T. Bidlake, now turned journalist, who was writing here – and in a cycling magazine. He was perhaps too immersed in the motor-traction bubble to understand the lord mayor's reaction. Indeed, it was not the lord mayor of Leeds that was the 'problem', but, rather, Bidlake's unpreparedness to presume anything less than a full embrace of motor traction must be myopic or backward. It is clear, though, that the lord mayor did miss a trick, and should have put his town first as so many motorized dignitaries worked their way through it on that day in May 1900. But his reaction – to feign disinterest – is perfectly understandable in the context of the time. The lord mayor derived status from the ceremony of his post, which included the spectacle of the procession of the horse-drawn carriage. Such a means of conveyance oozed discrimination and style. It spoke of certainties, permanence and dignity. It was what was expected of him as he undertook his duties of office. Associating with a motor-traction circus was a risk, and he cannot be 'blamed' for choosing not to take that risk, which could have placed him in the camp of an arrogant and wealthy elite, seeking to foist onto others their dangerous fad of the moment, demanding attention and road space, and leaving resentment and dust in their wake. Many more would probably have applauded the stance of the lord mayor.

---

[7] *Bicycling News and Motor Car Chronicle*, 2 May 1900, p. 19.
[8] *Bicycling News and Motor Car Chronicle*, 16 May 1900, p. 23.
[9] *Bicycling News and Motor Car Chronicle*, 16 May 1900, p. 20.

Anyway, an infrastructure needed to be put into place – availability of petrol, spares, repairs, hospitality. The road network was utterly unfit for its new use, with commentators suddenly finding that hills were 'dangerous' and signage inadequate; a debate raged over who should pay for their improvements.[10] As motor-car technology rapidly improved, public trials of motor cars moved away from merely testing hill-climbing and hill-starting ability – that is, matching and beating what a horse could do – and, instead, became much more demanding, leading ultimately to multi-thousand-mile tests, even driven entirely in top gear. Commentators were persuaded that the motor car had, by the First World War, reached a point where it was practically perfect and no further development could be made. Commentaries were by then speculating on the building of bypasses, or of special high-speed roads for motor cars only as part of a 'solution' for roads that were now clogged by motor traffic. The London to Brighton road was a case in point, now bunged up at weekends by that new breed of motorist, the 'steady and careful artisan', in search of the same dream of the 'open road' as the pioneers.

Motor traction had to be seen to be 'better' than the horse if it were to sell. It was not enough then for Bidlake, as a convert to the bicycle and motor car, to predict:

> In future ages when horses are no longer allowed to prance dangerously in public places, men will marvel at obsolete nineteenth-century notions. They will laugh at the idea of having a huge beast with a will of its own permitted to appear on the highway. They will marvel that the mind of man took such centuries to evolve mechanical locomotion and remained content with an attempt to coerce dangerous and powerful beasts to do their bidding.[11]

That was a view calculated to preach to the converted. Much more significant and useful for the undecided motorist were calculations published to 'prove' how cheap it was to run a motor car and emphasizing how keeping a horse required ongoing costs for feeding whether it was used or not. Commentators emphasized the distances that motor cars could cover, all day, every day, and which could not possibly be expected of the horse – Browne's journey, quite impossible by horse traction, was one of many to help build up this narrative. Edge wrote to *The Times* in 1905 pointing out the increasing unsuitability of the horse:

> I think if we had had mechanical traction before horse traction the idea of harnessing an animal like a horse and making him do work which in many instances must be continuous pain is cruelty of the most acute kind, and would be looked upon with horror by people today who take it as a matter of course.[12]

But even if the motor car was starting to create a demand for unfettered speed and distance, horse traction persisted. Thompson has shown that it was as late as

---

[10] See Plowden, *The Motor Car and Politics in Britain, 1896–1970*; Reid, *Roads Were Not Built for Cars*. See also Morrison and Minnis, *Carscapes*, pp. 239–43.
[11] *Bicycling News and Motor Car Chronicle*, 9 May 1900, p. 19.
[12] *The Times*, 4 July 1905.

1950 before the number of farm horses exceeded that of tractors.[13] While *Punch* traced a 'normalisation' of the motor car as a conveyance for the middle classes, and in place by the War, it also recorded the persistence of a horse-drawn world. Its cartoon, 'The survival of the fittest' in 1909, showed how 'only the willowy type is likely to survive the stress of modern traffic', in which a contorting woman finds herself in the middle of the street among mad delivery cyclists, a horse and cart and a horse and carriage, but not a single motorized vehicle.[14] Another cartoon of the same year featured two intertwined cyclists engaging in the 'rush and roar of our modern Babylon' (a high street at rush hour), completely engrossed in a newspaper. This was meant to emphasize the menace of the thoughtless cyclist, of course, but all of the drivers scattering to avoid them were in horse-drawn vehicles, with no motor cars to be seen.[15]

Cycling had become part of the everyday experience and by the War had long since ceased to be clubby, amateur and privileged. A swathe of cycling magazines covered every angle – racing, clothing, technology, gossip. The 'safety' had emerged, and fitted with pneumatic tyres, took cycling into the mainstream, ridden by men and women, young and old, rich and, well, not so rich. Cycling, though, had not so long before needed its champions, the riders enduring the abuse, missiles and dire roads, needing the strength in numbers brought by their club run, while 'scorchers' and women cyclists wearing 'rationals' had slowly massaged the attitudes and expectations of a wider society accustomed to a more pedestrian pace. The shocking was becoming the normal.

Even in the light of this kind of change, selling motor traction was quite a challenge for its entrepreneurs: how to persuade a public of the merits of motoring, an entirely novel pursuit and, frankly, a ridiculous one as well. But the entrepreneurs, for the most part, were drawing on their experience in bringing about the diffusion of the bicycle, also a metaphor for technological progress which had started off as equally irrelevant but which, with the evolution of the 'safety', had become entirely acceptable to its middle-class customer base; flick through any society magazine in the closing years of the nineteenth century, and articles abounded on the joys of cycling. Selling motor traction, then, required a combination of opportunities. It needed an enthusiastic hobbyist and newspaper press, and this drew on an existing and vibrant culture of cycling journalism to bring about not just motoring magazines such as *The Autocar* in 1895, which piqued a clear interest and encouraged other magazines in its wake, but a champion of motoring in Alfred Harmsworth, launching the hugely successful and pro-motoring *Daily Mail* in 1896. It needed friends in high places, and having a cycling royal household, then a motoring Prince of Wales, showing a keen interest, was a blessing. Motoring needed to be visible. This was already happening with motorists seen stricken at the roadside, or driving thoughtlessly through the village high street, their vehicles smoking, smelly and covered in mud. While those motorists appeared so often road-soiled and steeling themselves for abuse and missiles, it was all part of an image where motor traction slowly adjusted to become associated with reliability

---

[13] Thompson, 'Nineteenth-century horse sense', pp. 60–81, at p. 63.
[14] *Punch*, 3 February 1909, p. 77.
[15] *Punch*, 15 September 1909, p. 190.

and ease of use, and building further on the cyclists' discovery of freedom from train schedules, access to the 'open road' and enhanced leisure and business opportunities. And as with cycling, this freedom to roam then hinted at the further testing of the social boundaries of relationships and sexual licence, as well as the mobility, dress code and visibility of women. The seemingly innocent drawing by Charles Flower in a motoring magazine in 1902 depicted a young couple setting off on the 'open road' on a quadricycle, but that and the caption, 'An enjoyable experience: Their first trial trip'[16] hinted at so much more, hitherto impossible.

Motor traction needed to get beyond its initially small and cliquey customer base, with its clubs and customs that served to inform and support the right sort but exclude the many. It needed a confidence in the product, which would only come with its entrepreneurs' campaigning in the expensive and risky races and trials, and not just on home soil. Few were doing this but some entrepreneurs, Edge in particular, played the field, ensuring their products were photographed and reported to advantage. It needed the creation of new and trusted brands, and all the better if they were British. The emergence of a brand could happen quickly: Napier went from a precision-engineering company in the 1890s, with no motor-car or cycle heritage, to a race-proven high-quality British brand once Edge had won the all-important and internationally contested Gordon Bennett Cup in 1902. Motor traction also needed an information base, and this was already being provided in large part for cyclists by way of advice books, travel guides and maps with contours, indicating places of interest and beauty. Magazines soon offered alerts, updated weekly, of speed-trapped roads to watch out for. Also, clothing and accessories had been available in shops or by mail-order for cyclists and it was only natural to extend the range of stock offered, even if that meant simply adjusting the way items were promoted; books and maps, for example, once for the 'cyclist', could be reissued for the 'cyclist and motor tourist'. It needed motor vehicles to be, or at least to appear to be, affordable by a wider (middle-class) public, and articles and letters in the motoring press plus the occasional advice book assured the wavering that keeping a motor car or motor-cycle was much cheaper than keeping a horse. It needed a whiff of glamour, and entrepreneurs arranged for photogenic society and celebrity women to pose in driving seats for the covers of classy magazines such as *The Car Illustrated*. Edge's enterprises benefited greatly from the talented female driver Dorothy Levitt driving his Napiers in trials and races. It also helped the entrepreneur to broaden his scope, so while Edge paid lip service to aviation, he (and Levitt) campaigned for Napier motor-boats to great international effect.

It needed, then, entrepreneurs who could bring a track record of commitment and success, credibility and not a little stardust. Edge very much fitted this role, and he defended this status with aggressive letters to editors, with threats of legal recourse, by taking advantage of connections, and running the risks and costs that went with international competition. When he set up the Motor Power Company in 1899 he offered a full range of motor vehicles, from the Ariel motor-tricycle to the Napier, assuring his public that these had been 'selected as the best in their respective classes', having had

[16] *Motorcycling and Motoring*, 9 April 1902, p. 141.

Figure 8.1 Charles Flower drew this image, 'An enjoyable experience: Their first trial trip' for *Motorcycling and Motoring*. Many such drawings promised the open road and more for the courting couple: 9 April 1902, p. 141.

a perfectly free hand to deal only in those cars in which they [the directors] could have confidence. The public have thus the assurance that . . . they are securing the most reliable vehicle that is to be obtained, and are not experimenting at their own expense for the benefit of a maker who has nothing better to sell.[17]

---

[17] Motor Power Company brochure (1899), with thanks to the late Malcolm Jeal.

No doubt many other glossy brochures by other agents said much the same thing, but Edge followed his rhetoric up with 'successful' outcomes for all to see, from the victories in hill-climbs, trials and races, an association with the glamorous, wealthy and well connected, through to the endorsements from satisfied customers that he could then include in his brochures and adverts. Taking care to associate his Napiers with the best of British workmanship, meeting the challenge of the foreigner, he then followed this up by offering sales guarantees, and ultimately, Edge-branded accessories such as motor oils and tools.

Yet, the success of Napier and many other home-grown brands could not have happened in the longer term without the participation of a much broader social group of consumers, and that group was drawn in by the efforts of the entrepreneur. The investment in, and development of, the motor car could never have happened without an appeal beyond the clubby pioneers. For that insular group to continue to have it its way – treating motor traction as a sport for amateurs, or seeing little other purpose for the motor car beyond, say, taking guests to the shoot at the far end of their estate – would have been unsustainable. The 'Emancipation Run' of 1896, when, according to Brendon, the vehicles assembled outside the Metropole Hotel represented something like half of all the motor vehicles then in the country,[18] may be seen as a pivotal moment now, but in those days represented a fledgling industry that would have died without the risk, graft and investment of those first pioneering motoring entrepreneurs. Even by, say, 1899, comment on the use of the motor car in respectable society magazines was practically nil.

Edge could see this, and while he no doubt had one eye on the profits of his enterprises, he voiced an opinion as early as 1899 that his Automobile Club audience did not want to hear: motor traction had to broaden its social base. On the winding up of the Thousand Mile Trial in 1900, *The Times* saw benefits in this happening too. Once motor cars become cheaper, it said, more will sell, and England will benefit 'by the habitual use of an invention, . . . by circulating wealth and promoting activity in stagnant places, by improving roads, hotels, and all that is meant by communications, and by revealing to men the beauties of their native land'.[19] And so it was, since by the War many of the 'artisans' exploring such beauties did so on a motor-cycle, perhaps fitted with a side car, or even in a motor car or cyclecar, but without club membership. They could not have afforded club subscriptions, let alone club fripperies such as dinners or the right clothing, but then, they were not clubbable in the first place. A growing infrastructure, a second-hand market and updated maps all facilitated their weekend indulgence. Motoring very quickly became a broad church. The clubs retained their exclusive membership and benefits, but were no more able to resist 'the trade' contaminating their ranks any more than they could control the flow of new and increasingly socially unsuitable 'consumers' buying in to the game.

In time the 'old brigade' came to assume a venerable status. When the Circle of Nineteenth-Century Motorists was set up in 1927, Montague Grahame-White, one of its administrators, revealed that of the scores of applications, 'a very large percentage'

---

[18] Brendon, *The Motoring Century*, p. 13.
[19] 'The 1000-Mile trial of motor cars', *The Times*, 14 May 1900.

failed to pass the scrutiny of the committee;[20] this showed how real social capital came with membership. Edge was, of course, a member, and the dwindling clique attending its annual dinners into the 1930s read as a 'who's who' of the 'old brigade'. (Edge was also prominent in the Fellowship of Old-Time Cyclists, the table plan of whose twentieth annual dinner in 1929, for example, read like a 'Who's Who' of the early motoring world.)[21] He probably cherished membership of these old-timer groups, perhaps more so than the other trappings that followed his name: serving as president and benefactor for his cycling club, the Anerley Bicycle Club; being tapped for the occasional product endorsement; appearing on BBC radio; and cutting ribbons at openings. Indeed, and no doubt in part through his own efforts, he stayed in the public eye as he reached older age. In 1926, he was still sufficiently credible within the motor industry to be asked to endorse a film on road safety.[22] He was named frequently in the 1920s' *Illustrated London News* history feature, 'The Chronicle of the Car'.[23] He was an occasional subject of encomiums in the cycling press: for the new generation of ABC'ers in 1929, his feats were replayed. He cycled 'almost daily and we should not be surprised if he could even still sprint', his old cycling-club friend G. H. Smith wrote.[24] Edge remained good for the sound bite: his letter to *The Times* in 1933 displayed his customarily blunt opinion on the new driving tests (they were useless, and instead of tests, drivers should be more polite). He was keen to remind readers of his qualifications to pontificate on motoring matters, pointing out that he covered 30,000 miles a year, and by 1933 had driven one million miles.[25] Yet, 'I am a cyclist of over 45 years' experience', he wrote, and 'I still think it is the greatest health-giver there is. . . . A spin on a bicycle is an intense pleasure, and brings many happy days.'

So if motor traction could so nearly have been a curious nineteenth-century dead end, perhaps like the 'ordinary' bicycle, it could also have been just one more sporting hobby for the amateur to indulge in along with cycling, horse-driving, sailing and flying. The Lord Advocate, interviewed for *The Car Illustrated* in 1902, believed motor traction would supersede horse traction, but concluded that the two sports were quite distinct. Furthermore, as a means of exercise, he also thought cycling was 'without a rival', and motor-cycling simply 'an alternative source of enjoyment'. He took the view of 'many automobilists' that 'the sum of our pleasures is increased by the introduction of the petrol motor, but that it need not necessarily involve the supersession of other recreations. A man may still drive or ride a horse, or pedal a cycle, and yet indulge in some form of automobilism as well'.[26] In the same vein, cycling historian Nick Clayton

---

[20] Grahame-White, *At the Wheel Ashore and Afloat*, p. 239.
[21] 3 December 1929: the table plan shows where the 223 diners sat (Edge was at the top of the table). The plan for the following year, held on 3 December 1930, had 181 diners: University of Warwick, Modern Records Centre, MSS.328/N7/6/3/5.
[22] 'Hints to motorists on safety first by S. F. Edge' (1926), FHC Pictures in cooperation with the AA; British Film Institute ref. 18658.
[23] Edited variously by W. Whittall, and from 1927, John Prioleau and H. Thornton Rutter, this had been an ongoing feature of the *Illustrated London News* since 1907. In this version of popular motoring history, Edge played a pivotal role.
[24] Smith, *Some Notes about the Anerley B.C.*, p. 51. Edge joined the ABC in 1885.
[25] *The Times*, 26 August 1933; *Daily Mail*, 2 April 1931.
[26] C. L. Freeston, 'The Lord Advocate on motor-cycling', *The Car Illustrated*, 11 June 1902, p. 79.

"FELLOWSHIP OF OLD TIME CYCLISTS", 1927; 24 CATFORD C.C. MEMBERS PRESENT.

Figure 8.2 The annual dinner of the Fellowship of Old-Time Cyclists, 1927. S. F. Edge (right) presides over a table. Image: Catford Cycling Club, from E. J. Southcott (ed.), *The first fifty years of the Catford Cycling Club* (London: G. T. Foulis, 1939), facing p. 241.

has pondered how riding the 'ordinary' might have survived as a niche sport – alongside the 'safety' as a classless activity for all – had better roads come sooner than they did.[27]

Instead, motor traction gained a critical mass such that by about 1905 magazine editors and entrepreneurs were looking back and reminding readers how they were the visionary ones – but what exactly was their vision? It was, from one perspective, going to be a motorized future for the few, serving their convenience and leisure. But other views, particularly those of Edmund Dangerfield and *The Motor*, saw a much wider engagement, bringing opportunities for enhanced leisure and business. That magazine's crusade to create the 'modest motorist', and with its spin-off magazine *The Cyclecar* in creating the cyclecarist, would appear to have been remarkably successful; the number of registered vehicles in 1904 was 8,000; by 1914 it was 132,000.[28] The road system could not cope. This had been achieved by the creation of a 'mature' product: a light car which, by about 1905, had a standardized 'look' and an understood mechanical layout. Admittedly, there was little that was 'standardized' about the cyclecar breed, but its consumers were joining a motor-traction community accustomed to seeking the 'open road' and using an infrastructure now sufficiently in place to do so. There was now a tolerable reliability, and improvements in tyres meant drives were plagued less by puncturing; where *The Motor* had urged 'modest motorists' to do their own maintenance to keep costs down, this advice lapsed as reliability improved. Petrol-fuelled internal combustion was the clear winner, and whatever its merits compared to electric or steam combustion, this started to offer some certainty to the buyer. The motor car by the First World War was increasingly bought second-hand or on tick by

---

[27] Clayton, 'The birth of Tarmacadam', pp. 88–92, at p. 92.
[28] B. R. Mitchell, *Abstract of British Historical Statistics* (Cambridge: Cambridge University Press, 1962), p. 230.

a new, modest-income buyer who had less expectation, or inclination, to tinker and toil at the roadside. The pioneering days (up to about 1902/3) were increasingly looked back on with amusement, even horror, and Edwardian writers wondered just how the motor car could now possibly get any better. This book started in the 1890s with the attention-drawing spectacle of the motorized advertising hoarding. Then, people stared at the motor car as much as at the hoarding. By the First World War those same people had long since stopped noticing the motor car.

Bibliography

Primary sources

Key periodicals

The Autocar
Automotor and Horseless Vehicle Journal
Bicycling News and Motor Car Chronicle
The Car Illustrated
The Cyclecar (*The Light Car and Cyclecar*)
Cycling
Daily Mail
Harmsworth London Magazine
The Light Car
Motorcycling and Motoring (*The Motor*)
Motor Car Journal
The Motor Cycle
Motorcycling
Motoring Illustrated
Penny Illustrated Paper
Punch
The Strand
The Times

Unpublished material

ACQ4, 'The Circle of Nineteenth-Century Motorists', RAC archive
ACQ 9/1, Alfred Bradley diaries, RAC archive
Horace Mansell, 'An imperfect account of a "Safety" cycling tour taken during my summer's holiday, August-September 1888', unpub. travelogue, 5 vols (private collection)
Edge scrapbooks, vol. 141, Veteran Car Club archive
MSS.328, University of Warwick, Modern Records Centre, various
Nixon/1, loose papers, National Motor Museum library, Beaulieu
Science and Industry Museum Archive, Manchester, various
Unpublished diary of G. H. Smith for the period 1891–6, consulted with the kind permission of Penny Morris

Published material

The book of Edge accessories for motor cars (1912).
[anon.], 'Motors on tour: A holiday with the Automobile Club', *The Daily News*, 13 April 1898.
[anon.], *The Black Anfielders Being the Story of the Anfield Bicycle Club, 1879–1955* ([Liverpool]: Anfield Bicycle Club, 1956).
[anon.], *The British Motor Tourist A.B.C.* (London: The British motor tourist A. B. C. Co., 1905) (foreword by S. F. Edge).
[anon.], *The Hooley Book: The Amazing Financier, His Career and His Crowd* (London: J. Dicks, 1904).
[anon.], *Chauffeur's Blue Book* (1906), repr.: *Etiquette for the Chauffeur: Duties and Hints for Chauffeurs* (East Grinstead: Copper Beech, 1997).
N. G. Bacon, 'Modest motoring', *The Girl's Own* Paper, 5 December 1903, pp. 148–50.
[His Grace the Duke of] Beaufort, *Driving* (London: Longmans, Green & Co., 4th ed., 1894).
A. E. Berriman, *Motoring: An Introduction to the Car and the Art of Driving It* (London: Methuen & Co, 1914).
W. C. A. Blew, 'The coaching revival', in [His Grace the Duke of] Beaufort, *Driving* (London: Longmans, Green & Co., 4th ed., 1894), pp. 273–305.
H. Massac Buist (ed.), *The Motor Year Book and Automobilist's Annual 1906* (London: Methuen, 1906).
H. Massac Buist, *Motoring to Stonehenge* (London: Anglo-American Oil Company, 1907).
Viscount Bury and G. Lacy Hillier, *Cycling* (London: Longmans, Green, & Co, 3rd ed., 1891).
Marquis de Chasseloup-Laubat, 'A short history of the motor-car', in [Lord] Northcliffe (ed.), *Motors and Motor Driving* (London: Longmans, Green and Co., 4th ed., 1906), pp. 1–24.
John Dillon, *Motor Days in England: A Record of a Journey Through Picturesque Southern England with Historical and Literary Observations by the Way* (London: G. P. Putnam's Sons, 1908).
Major Henry Dixon, 'Hints to beginners, Part I', in [His Grace the Duke of] Beaufort, *Driving* (London: Longmans, Green & Co., 4th ed., 1894), pp. 116–30.
S. F. Edge, 'Motor cars and modern warfare', in Roger Pocock, *The Frontiersman's Pocket-Book* (1909; Edmonton, AB: University of Alberta Press, 2012), pp. 223–6.
S. F. Edge and Charles Jarrott, 'Motor driving', in Alfred Harmsworth (ed.), *Motors and Motor Driving* (London: Longmans, Green & Co, 1902), pp. 322–40.
F. J. Erskine, *Lady Cycling: What to Wear and How to Ride* (1897; London: British Library Publishing, 2014).
Alexander Bell Filson Young, *The Complete Motorist* (London: Methuen, 1904).
Alexander Bell Filson Young, *The Happy Motorist: An Introduction to the Use and Enjoyment of the Motor Car* (London: E.G. Richards, 1906).
C. L. Freeston, 'Automobile literature', in Alfred Harmsworth (ed.), *Motors and Motor Driving* (London: Longmans, Green and Co., 1902), pp. 397–401.
W. J. Gordon, *The Horse World of London* (London: The Religious Tract Society, 1893).
Montague Grahame-White, *At the Wheel Ashore and Afloat* (London: G.T. Foulis, n.d. [*c*. 1935]).
Alfred Harmsworth (ed.), *Motors and Motor Driving* (London: Longmans, Green & Co, 1902).

James John Hissey, *Untravelled England* (London: Macmillan, 1906).
J. J. Hissey, *A Holiday on the Road: An Artist's Wanderings in Kent, Sussex and Surrey* (London: Richard Bentley and Son, 1887).
E. T. Hooley, *Hooley's Confessions* (London: Simpkin, Marshall, Hamilton and Kent, n.d. [1924]).
Charles Howard, *The Roads of England and Wales: An Itinerary for Cyclists, Tourists and Travellers* (1882; London: George Gill and Sons, 5th ed., 1897).
Charles Jarrott, *Ten Years of Motors and Motor Racing* (London: Motor Sport, 1929).
W. Rees Jeffreys, 'The Motor Union of Great Britain and Ireland and its work', in [Lord] Northcliffe (ed.), *Motors and Motor Driving* (London: Longmans, Green and Co., 4th ed., 1906), pp. 420–8.
[Lady] Jeune, 'Dress for motoring. I. Dress for ladies', in Alfred Harmsworth (ed.), *Motors and Motor Driving* (London: Longmans, Green & Co, 1902), pp. 66–71.
Mary Eliza Kennard, *The Motor Maniac* (London: Hutchinson and Co, 1902).
Dorothy Levitt, *The Woman and Her Car: A Chatty Little Book for Women Who Motor or Want To Motor* (1909; London: Hugh Evelyn, 1970).
Owen Llewellyn and L. Raven-Hill, *The South-Bound Car* (London: Methuen, 1907).
Major C. G. Matson, *The Modest Man's Motor* (London: Lawrence and Bullen Ltd, 1903).
John Scott-Montagu, *The First Ten Years of Automobilism, 1896–1906* (London: The Car Illustrated, 1906).
John Scott-Montagu, 'The utility of motor vehicles', in Alfred Harmsworth (ed.), *Motors and Motor Driving* (London: Longmans, Green & Co, 1902).
[Lord] Northcliffe (ed.), *Motors and Motor Driving* (London: Longmans, Green and Co., 4th ed., 1906).
Max Pemberton, *The Amateur Motorist* (London: Hutchinson and Co, 1907).
Gerald Rose, *A Record of Motor Racing, 1894–1908* (1909; London: Motor Racing Publications, 1949).
G. H. Smith, *Selwyn Francis Edge: The Man and Some of the Things He Has Done* (privately pub., 1928).
[G. H. Smith], *Some Notes about the Anerley B.C.* (privately pub., 1930).
Colonel Hugh Smith-Baillie, 'Hints to beginners, Part II', in [His Grace the Duke of] Beaufort, *Driving* (London: Longmans, Green & Co., 4th ed., 1894), pp. 131–7.
Henry Sturmey, *On an Autocar Through the Length and Breadth of the Land* (London: Illife, Sons and Sturmey, 1898).
Henry R. Sutphen, 'Touring in automobiles', *Outing*, 38:3 (June 1901), pp. 197–202.
J. E. Vincent, *Through East Anglia in a Motor Car* (London: Methuen, 1907).
A. J. Wilson (ed.), *Motor Trips at a Glance: In England, Scotland, Wales, Ireland and France* (London: A. J. Wilson & Co., 1911).

Secondary sources

Daryl Adair, 'Spectacles of speed and endurance: The formative years of motor racing in Europe', in David Thoms, Len Holden and Tim Claydon (eds), *The Motor Car and Popular Culture in the Twentieth Century* (Aldershot: Ashgate, 1998), pp. 120–34.
R. J. Q. Adams, *Balfour: The Last Grandee* (London: John Murray, 2007).
Martin Adeney, *The Motor Makers* (London: HarperCollins, 1988).

Roger Alma, *The Malvern Cycling Club, 1883-1912: A History of Cycling in Malvern, Worcestershire, before the First World War* (John Pinkerton Memorial Publishing Fund, 2011).
Thomas Ameye, Bieke Gils and Pascal Delheye, 'Daredevils and early birds: Belgian pioneers in automobile racing and aerial sports during the Belle Epoque', *International Journal of the History of Sport*, 28:2 (2011), pp. 205-39.
Shima Amini and Steven Toms, 'Accessing capital markets: Aristocrats and new share issues in the British bicycle boom of the 1890s', *Business History*, 60:2 (2018), pp. 231-56.
John Archer, 'The Manchester Automobile Club Reliability Run, 1907-14', *Aspects of Motoring History*, 13 (2017), pp. 49-70.
John E. Archer, '"Men behaving badly?": Masculinity and the uses of violence, 1850-1900', in Shani D'Cruze (ed.), *Everyday Violence in Britain, 1850-1950* (Harlow: Longman, 2000), pp. 41-54.
Arthur C. Armstrong, *Bouverie Street to Bowling Green Lane: Fifty-five Years of Specialized Publishing* (London: Hodder, 1946).
T. S. Ashton, 'Business history', in Richard Davenport-Hines and Peter Treadwell (eds), *Capital, Entrepreneurs and Profits* (Abingdon: Frank Cass, 1990).
Chris Barker, 'The rise and fall of the Model T in Britain', *Aspects of Motoring History*, 12 (2016), pp. 33-48.
Chris Barker, Neil Tuckett and Drew Lilleker, *The English Model T Ford, Vol. 2: Beyond the Factory* (Keighley: Model T Ford Register of Great Britain, 2014).
Peter Barlow and Martin Boothman (eds), *'Conspicuously Marked': Vehicle Registration in Gloucestershire, 1903-13* (Gloucester: Bristol & Gloucestershire Archaeological Society, 2019).
T. M. Barlow and J. Fletcher (comps.), *A History of Manchester Wheelers' Club, 1883-1983* (Manchester: 1983).
Hugh Barty-King, *The AA: A History of the Automobile Association, 1905-1980* (Basingstoke: Automobile Association, 1980).
Bradley J. Beaven, 'The growth and significance of the Coventry car component industry, 1895-1939' (unpub. PhD dissertation, De Montfort University, 1995).
Elizabeth Bennett, *Thousand Mile Trial* (Heathfield: Elizabeth Bennett, 2000).
Geraldine Biddle-Perry, 'The rise of "The world's largest sport and athletic outfitter": A study of Gamage's of Holborn, 1878-1913', *Sport in History*, 34:2 (2014), pp. 295-317.
Geraldine Biddle-Perry, 'Fashioning suburban aspiration: Awheel with the Catford Cycling Club, 1886-1900', *London Journal*, 39:3 (2014), pp. 1-18.
Wiebe Bijker, *Of Bicycles, Bakelites, and Bulbs: Toward a Theory of Sociotechnical Change* (Cambridge, MA: MIT Press, 1995).
Anthony Bird, *The Motor Car, 1765-1914* (London: B. T. Batsford, 1960).
Roger Bird, *The Birth of Brooklands* (Woking: privately pub., 2012).
David Birchall, 'George Pilkington Mills and the first International Bordeaux-Paris race', *The Boneshaker*, 197 (2015), pp. 12-19.
W. Boddy, *The Story of Brooklands: The World's First Motor Course, Vol. 1* (London: Grenville, 1948).
John Bolster, *Motoring is My Business* (London: Autosport, 1958).
Ian Boutle, '"Speed lies in the lap of the English": motor records, masculinity and the nation 1907-14', *Twentieth-Century British History*, 23:4 (2012), pp. 449-72.
W. F. Bradley, *Motor Racing Memories* (London: Motor Racing Publications, 1960).
Ruth Brandon, *Automobile: How the Car Changed Life* (London: Macmillan, 2002).

Piers Brendon, *The Motoring Century: The Story of the Royal Automobile Club* (London: Bloomsbury/RAC, 1997).
Christopher Breward, *The Hidden Consumer: Masculinities, Fashion and City Life, 1860-1914* (Manchester: Manchester University Press, 1999), pp. 242-6.
Gordon Brooks, 'The Small Car Trials of 1904', *Aspects of Motoring History*, 3 (2007), pp. 18-39.
Barbara Burman, 'Racing bodies: Dress and pioneer women aviators and racing drivers', *Women's History Review*, 9:2 (2000), pp. 299-326.
Frank Hedges Butler, *Fifty Years of Travel by Land, Water, and Air* (London: T. Fisher Unwin Ltd, 1920).
Josh Butt, 'The diffusion of the automobile in the North-West of England, 1896-1939' (unpub. PhD dissertation, Manchester Metropolitan University, 2019).
Josh Butt, 'Adapting to the emergence of the automobile: A case study of Manchester coachbuilder Joseph Cockshoot and Co., 1896-1939', *Science Museum Group Journal*, 8 (2017) [no pp].
James Buzard, *The Beaten Track: European Tourism, Literature and the Ways to Culture, 1800-1918* (Oxford: Clarendon Press, 1993).
C. F. Caunter, *The Light Car: A Technical History* (London: HMSO/The Science Museum, 1970).
Roy Church, *The Rise and Decline of the British Motor Industry: Studies in Economic and Social History* (London: Macmillan, 1994).
Tom Clarke, 'The first motor-car journey in England: A driving controversy', *Aspects of Motoring History*, 12 (2016), pp. 67-81.
Georgine Clarsen, *Eat My Dust: Early Women Motorists* (Baltimore, MD: Johns Hopkins University Press, 2008).
Anders Ditlev Clausager, *Wolseley: A Very British Car* (Beaworthy: Herridge and Sons, 2016).
Nick Clayton, 'The splendid bankrupt', paper delivered at the 30th International Cycling History Conference, Znojmo, Czechia, 18-22 June 2019.
Nick Clayton, 'Le baptême de la bicyclette', *The Boneshaker*, 203 (2017), pp. 18-22.
Nick Clayton, 'The birth of Tarmacadam', *Cycle History*, 24 (2013), pp. 88-92.
Nick Clayton, 'A missed opportunity? Bicycle manufacturing in Manchester, 1880-1900', in Derek Brumhead and Terry Wyke (eds), *Moving Manchester: Aspects of the History of Transport in the City and Region Since 1700* (Manchester: Lancashire and Cheshire Antiquarian Society, 2004).
Tracy Collins, 'Athletic fashion, "Punch", and the creation of the New Woman', *Victorian Periodical Review*, 43 (2010), pp. 309-35.
Marilyn Constanzo, '"One can't shake off the women": Images of sport and gender in *Punch*, 1901-10', *International Journal of the History of Sport*, 19:1 (2002), pp. 31-56.
Esme Coulbert, 'Perspectives on the road: Narratives of motoring in Britain, 1896-1930' (unpub. PhD dissertation, Nottingham Trent University, 2013).
Esme Coulbert, '"The romance of the road": Narratives of motoring in England, 1896-1930' in Benjamin Colbert (ed.), *Travel Writing and Tourism in Britain and Ireland* (Basingstoke: Palgrave Macmillan, 2012), pp. 201-18.
Donald Cowbourne, *British Trial Drivers: Their Cars and Awards, 1902-1914* (Otley: Westbury Publishing, 2003).
S. C. H. Davis, *Atalanta: Women as Racing Drivers* (London: G. T. Foulis & Co, n.d. [c. 1950]).
S. C. H. Davis, *Motor Racing* (London: Iliffe & Sons, 1932).

Graham Dawson, *Soldier Heroes: British Adventure, Empire and the Imagining of Masculinities* (London: Routledge, 1994).
Andrew Denning, 'Transports of speed', in Michael Saler (ed.), *The Fin-de-siècle World* (London: Routledge, 2015).
P. E. Dewey, *British Agriculture in the First World War* (London: Routledge, 1989).
Colin Divall, 'Mobilizing the history of technology', *Technology and Culture*, 51:4 (2010), pp. 938–60.
Hugh Driver, *The Birth of Military Aviation: Britain, 1903–1914* (Woodbridge: Boydell Press, 1997).
[Sir] Arthur Du Cros, *Wheels of Fortune: A Salute to Pioneers* (London: Chapman and Hall Ltd, 1938).
Enda Duffy, *The Speed Handbook: Velocity, Pleasure, Modernism* (Durham, NC: Duke University Press, 2009).
H. O. Duncan, *The World on Wheels*, Vol. 2 (Paris: privately pub., 1926).
Tim Edensor, 'Automobility and national identity: Representation, geography and driving practice', *Theory, Culture and Society*, 21:4–5 (2004), pp. 101–20.
Selwyn Francis Edge, *My Motoring Reminiscences* (London: G. T. Foulis, 1934).
S. F. Edge, 'Looking back: Interesting comparisons of yesterday and today by a pioneer', *Morris Overseas Mail*, May 1931, pp. 7–9.
David Edgerton, *Science, Technology and British Industrial 'Decline', 1870–1970* (Cambridge: Cambridge University Press, 1996).
Michael Edwards, *The Tricycle Book, 1895–1902, Part One* (Brighton: Surrenden Press, 2018).
Michael Edwards, *The Tricycle Book, 1895–1902, Part Two* (Brighton: Surrenden Press, 2019).
Michael Edwards, 'The Circle of Nineteenth-Century Motorists', *The Automobile*, 23:10 (2005), pp. 48–50.
Hubert W. Egerton (ed., Malcolm Jeal), 'Early motoring experiences', *Aspects of Motoring History* 4 (2008), pp. 4–17.
Roger Ellesmere, *British Caravans, Vol. 1: Makes Founded Before World War II* (Beaworthy: Herridge and Sons, 2012).
Clive Emsley, '"Mother, what did policemen do when there weren't any motors?": the law, the police and the regulation of motor traffic in England, 1900–1939', *Historical Journal*, 36:2 (1993), pp. 357–81.
J. M. Fenster, 'Mr Jarrott and Mr Edge: An Edwardian rivalry', *Automobile Quarterly*, 29:1 (1991), pp. 96–105.
David Fisher, *Cinema-by-Sea: Film and Cinema in Brighton and Hove since 1896* (Hove: Terra Media, 2012).
Simon Fisher, 'History: S. F. Edge (1907) Ltd', *Napier Heritage News*, 95 (2017), pp. 6–7.
Richard Fletcher, 'Charles Lincoln Freeston (1865–1942): Transport writer and journalist', *Aspects of Motoring History*, 14 (2018), pp. 29–40.
James J. Flink, *America Adopts the Automobile, 1895–1910* (Cambridge, MA: MIT Press, 1970).
James J. Flink, *The Automobile Age* (Cambridge, MA: MIT Press, 1988).
Raymond Flower and Michael Wynn Jones, *One Hundred Years of Motoring: An RAC Social History of the Car* (London: RAC Publishing, 1981).
James Foreman-Peck, Sue Bowden and Alan McKinlay, *The British Motor Industry* (Manchester: Manchester University Press, 1995).
Kathleen Franz, *Tinkering: Consumers Reinvent the Early Automobile* (Philadelphia, PA: University of Pennsylvania Press, 2005).

Nicholas Freeman, *1895: Drama, Disaster, and Disgrace in Late Victorian Britain* (Edinburgh: Edinburgh University Press, 2011).
Michael French and Jim Phillips, *Cheated Not Poisoned? Food Regulation in the United Kingdom, 1875–1938* (Manchester: Manchester University Press, 2009).
Ian Gazeley, 'Income and living standards, 1870–2010', in Roderick Floud, Jane Humphries and Paul Johnson (eds), *The Cambridge Economic History of Modern Britain, Vol. 2: 1870 to the Present* (Cambridge: Cambridge University Press, 2014).
Steven M. Gelber, *Horse Trading in the Age of Cars: Men in the Marketplace* (Baltimore, MD: Johns Hopkins University Press, 2008).
G. N. Georgano (ed. in chief), *The Beaulieu Encyclopedia of the Automobile: Coachbuilding* (London: The Stationery Office, 2001).
G. N. Georgano (ed. in chief), *The Beaulieu Encyclopedia of the Automobile* (London: The Stationery Office, 3 vols, 2000).
G. N. Georgano, *Cars, 1886–1930* (Gothenburg: AB Nordbok, 1985).
Nick Georgano (ed.), *Britain's Motor Industry* (London: G. T. Foulis and Co., 1995).
B. Gibson, 'From the charabanc to the gay hostess', in Robert Snape and Daniel Smith (eds), *Recording Leisure Lives: Holidays and Tourism in Twentieth-Century Britain* (Eastbourne: Leisure Studies Association, 2011).
Bryan Goodman, *American Cars in Prewar England: A Pictorial Survey* (Jefferson, NC: McFarland, 2004).
T. R. Gourvish, *Railways and the British Economy, 1830–1914* (London: Palgrave Macmillan, 1980).
Tony Hadland, *The Sturmey-Archer Story* (n.p.: Tony Hadland, 1987).
H. Hazel Hahn, 'Consumer culture and advertising', in Michael Saler (ed.), *The Fin-de-siècle World* (Abingdon, New York: Routledge, 2015), pp. 392–408.
Charles G. Harper, *The Brighton Road: Speed, Sport and History on the Classic Highway* (London: Chapman and Hall Ltd, 1906).
Martin Harper, *Mr Lionel: An Edwardian Episode* (London: Cassell, 1970).
Paul Harris, *Life in a Scottish Country House: The Story of A. J. Balfour and Whittingehame House* (Whittingehame: Whittingehame House Publishers, 1989).
A. E. Harrison, 'The competitiveness of the British cycle industry, 1890–1914', *Economic History Review*, 22:2 (1969), pp. 287–303.
Anthony Edward Harrison, 'Growth, entrepreneurship and capital formation in the UK cycle and related industries, 1870–1914' (unpub. PhD dissertation, University of York, 1977).
A. E. Harrison, 'Joint-stock company flotation in the cycle, motor vehicle and related industries, 1882–1914', *Business History*, 23 (1981), pp. 165–90, repub. as A. E. Harrison, 'Joint-stock company flotation in the cycle, motor-vehicle and related industries, 1882–1914', in Richard Davenport-Hines and Peter Treadwell (eds), *Capital, Entrepreneurs and Profits* (Abingdon: Frank Cass, 1990), pp. 206–32.
Adrian Harvey, *The Beginnings of a Commercial Sporting Culture in Britain, 1793–1850* (Aldershot: Ashgate, 2004).
Oliver Heal, 'Sir Ambrose Heal and the automobile', *Aspects of Motoring History*, 12 (2016), pp. 100–102.
John Heitmann, *The Automobile and American Life* (Jefferson, NC: McFarland, 2009).
John Hibbs, *The History of British Bus Services* (Newton Abbot: David and Charles, 1968).
Ian Hicks (ed.), *Early Motor Vehicle Registration in Wiltshire, 1903–1914* (Trowbridge: Wiltshire Record Society, 2006).

Craig Horner, 'S. F. Edge: The salad days', *Aspects of Motoring History*, 11 (2015), pp. 43–54.
Craig Horner, *The Cheshire Motor Vehicle Registrations, Vol. 1: 1904–07* (Liverpool: Record Society of Lancashire and Cheshire, 2019).
Craig Horner, 'The emergence of automobility in the United Kingdom', *Transfers*, 2:3 (2012), pp. 56–75.
Stephen Howarth, *A Century in Oil: The 'Shell' Transport and Trading Company, 1897–1987* (London: Weidenfeld and Nicholson, 1997).
Alun Howkins, *The Death of Rural England: A Social History of the Countryside* (London: Routledge, 2003).
Mike Huggins, 'Sport and the British upper classes, *c*. 1500–2000: A historiographic overview', *Sport in History*, 28 (2008), pp. 364–88.
Harry Irwin, *The Hum of the Wheels, the Roar of the Crowd: Five Cycling Shaw Brothers from Woolloomooloo* (Melbourne: Australian Scholarly Publishing, 2018).
Chris Ivory and Audley Genus, 'Symbolic consumption signification and the "lock-out" of electric cars, 1885–1914', *Business History*, 52:7 (2010), pp. 1107–22.
'Ixion' [Canon B. H. Davies], *Motor Cycle Cavalcade* (London: Iliffe and Sons, 1950).
Malcolm Jeal, 'Facts spoil a bloody tale: The 1903 Paris-Madrid Race', *Aspects of Motoring History*, 1 (2005), pp. 32–5.
Malcolm Jeal (ed.), *London to Brighton Run (Centenary)* (Crawley: Consortium, 1996).
Malcolm Jeal, 'Transport and travel in 19th-century Britain', in Malcolm Jeal (ed.), *London to Brighton Run (Centenary)* (Crawley: Consortium, 1996), pp. 8–12.
Malcolm Jeal, 'Pioneer British motor car makers', in Malcolm Jeal (ed.), *London to Brighton Run (Centenary)* (Crawley: Consortium, 1996), pp. 59–63.
Malcolm Jeal, 'By and Large: Napier – Britain's first purpose-built racing car in context', *Aspects of Motoring History*, 5 (2009), pp. 36–64.
David Jeremiah, 'Motoring and the British countryside', *Rural History*, 21:2 (2010), pp. 233–50.
David Jeremiah, 'Filling up: The British experience, 1896–1940', *Journal of Design History*, 8:2 (1995), pp. 97–116.
David Jeremiah, *Representations of British Motoring* (Manchester: Manchester University Press, 2007).
Bob Johnston and Derek Stuart-Findlay, *The Motorist's Paradise: An Illustrated History of Early Motoring in and Around Cape Town* (Cape Town: Bob Johnston and Derek Stuart-Findlay, 2007).
Kat Jungnickel, *Bikes and Bloomers: Victorian Women Inventors and Their Extraordinary Cycle Wear* (London: Goldsmiths Press, 2018).
Katrina Jungnickel, '"One needs to be very brave to stand all that": Cycling, rational dress and the struggle for citizenship in late nineteenth-century Britain', *Geoforum*, 64 (2015), pp. 362–71.
David Keir and Bryan Morgan (eds), *Golden Milestone: 50 Years of the AA* (London: Automobile Association, 1955).
Stephen Kern, *The Culture of Time and Space, 1880–1918* (Cambridge, MA: Harvard University Press, 1983).
Steve Koerner, 'The British motor cycle industry during the 1930s', *Journal of Transport History*, 16:1 (1995), pp. 55–76.
James M. Laux, *In First Gear: The French Automobile Industry to 1914* (Liverpool: Liverpool University Press, 1976).

Michael John Law, *The Experience of Suburban Modernity: How Private Transport Changed Interwar London* (Manchester: Manchester University Press, 2014).
Michael John Law, 'Driving to the "Super" Roadhouse', *Aspects of Motoring History*, 12 (2016), pp. 49–60.
Michael John Law, 'Charabancs and social class in 1930s' Britain', *Journal of Transport History*, 36:1 (2015), pp. 41–57.
Zoe Lawson, 'Wheels within wheels: The Lancashire cycling clubs of the 1880s and 1890s', in A. G. Crosby (ed.), *Lancashire Local Studies in Honour of Diana Winterbotham* (Lancaster: Carnegie Publishing, 1993), pp. 123–45.
Keith Laybourn with David Taylor, *The Battle for the Roads of Britain: Police, Motorists and the Law, c. 1890s to 1970s* (London: Palgrave McMillan, 2015).
Sally Ledger and Roger Luckhurst (eds), *The Fin de Siècle: A Reader in Cultural History, c. 1880–1900* (Oxford: Oxford University Press, 2000).
Edward Liveing, *Pioneers of Petrol: A Centenary History of Carless, Capel and Leonard, 1859–1959* (London: H. F. and G. Witherby Ltd., 1959).
Roger Lloyd-Jones and M. J. Lewis with the assistance of M. Eason, *Raleigh and the British Bicycle Industry: An Economic and Business History, 1870–1960* (Aldershot: Ashgate, 2000).
Stan Lockwood, *Kaleidoscope of Char-a-bancs and Coaches* (London: Marshall, Harris and Baldwin, n.d. [c. 1980]).
Sam Lomax and John Norris, *Early Days: Memories of the Beginnings of Automobile Engineering in South Lancashire and Cheshire* (Manchester: privately published, c. 1946).
Eilidh Macrae, 'The Scottish cyclist and the new woman: Representations of female cyclists in Scotland, 1890–1914', *Journal of Scottish Historical Studies*, 35:1 (2015), pp. 70–91.
D. McCullough, *The Wright Brothers: The Dramatic Story Behind the Legacy* (London: Simon & Shuster, 2015).
Chip McGoun, 'Automobile commerce and competition in the nineteenth century', unpub. paper, delivered at the fourth Annual Michael Argetsinger Symposium, Watkins Glen, 9 November 2018.
Clay McShane, *The Horse in the City: Living Machines in the Nineteenth Century* (Baltimore, MD: Johns Hopkins University Press, 2011).
Clay McShane, *Down the Asphalt Path: The Automobile and the American City* (New York: Columbia University Press, 1994).
Philip Gordon Mackintosh and Glen Norcliffe, 'Men, women and the bicycle: Gender and the social geography of cycling in the late nineteenth century', in Dave Horton, Paul Rosen and Peter Cox (eds), *Cycling and Society* (Aldershot: Ashgate, 2011), pp. 153–77.
Susan Major, *Early Victorian Railway Excursions: The Million Go Forth* (Barnsley: Pen and Sword, 2016).
David Matless, *Landscape and Englishness* (London: Reaktion Books, 1998).
Christoph Maria Merki, 'The birth of motoring out of sport: Car racing as a public relations strategy, 1894–1905', in Laurent Tissot and Béatrice Veyrassat (eds), *Technological Trajectories, Markets, Institutions: Industrialized Countries, 19th–20th Centuries* (Bern: Peter Lang, 2001), pp. 227–249.
Andrew Miles, 'Social structure, 1900–1939', in Chris Wrigley (ed.), *A Companion to Early Twentieth-Century Britain* (London: Blackwell, 2003).
Henry Miller, 'The problem with *Punch*', *Historical Research*, 82:216 (2009), pp. 285–302.
Nigel J. G. Y. Mills, 'Henry Lawson and the birth of the British motor industry: "A matter of opinion"' (unpub. MPhil dissertation, De Montfort University, 2002).

John Minnis, 'Sir David Solomons's motor stables, Broomhill, Southborough, Tunbridge Wells, Kent' (English Heritage Research Department Report Series, No. 7, 2009).
John Minnis, *England's Motoring Heritage from the Air* (London: Historic England Publishing, 2014).
Gijs Mom, *Atlantic Automobilism: The Emergence and Persistence of the Car, 1895–1940* (New York: Berghahn, 2014).
Gijs Mom, *The Electric Vehicle: Technology and Expectations in the Automobile Age* (Baltimore, MD: Johns Hopkins University Press, 2005).
Gijs Mom, 'Civilised adventure as a remedy for nervous times': Early automobilism and fin de siècle culture', *History of Technology*, 23 (2001), pp. 157–90.
[Lord] Montagu of Beaulieu and David Burgess-Wise, *A Daimler century: The full history of Britain's oldest car maker* (Sparkford: Patrick Stephens, 1995).
[Lord] Montagu of Beaulieu and Patrick Macnaghten, *Home James: The Chauffeur in the Golden Age of Motoring* (London: Wiedenfeld and Nicholson, 1982).
[Lord] Montagu of Beaulieu and Michael Sedgwick, *The Gordon Bennett Races* (London: Cassell, 1965).
Massimo Moraglio, 'Knights of Death: Introducing bicycles and motor vehicles to Turin, 1890–1907', *Technology and Culture*, 56:2 (2015), pp. 370–93.
Steven Morewood, *Pioneers and Inheritors: Top Management in the Coventry Motor Industry, 1896–1972* (Coventry: Centre for Business History, Coventry Polytechnic, 1990).
Simon Morgan, 'Celebrity: Academic "pseudo-event" or a useful concept for historians?', *Cultural and Social History*, 8 (2011), pp. 95–114.
Kathryn A. Morrison and John Minnis, *Carscapes: The Motor Car, Architecture and Landscape in England* (New Haven, CT: Yale University Press, 2012).
Kurt Möser, 'The dark side of automobilism, 1900–30: Violence, war and the motor car', *Journal of Transport History*, 24:2 (2003), pp. 238–58.
Kurt Möser, 'World War I and the creation of desire for automobiles in Germany', in Susan Strasser, Charles McGovern and Matthias Judt (eds), *Getting and Spending: European and American Consumer Societies in the Twentieth Century* (Cambridge: Cambridge University Press, 1998), pp. 195–222.
Joe Moran, *On Roads: A Hidden History* (London: Profile, 2009).
Roger Munting, 'The games ethic and industrial capitalism before 1914: The provision of company sports', *Sport in History*, 23 (2003), pp. 45–63.
T. R. Nicholson, *Sprint: Speed Hillclimbs and Speed Trials in Britain, 1899–1925* (Newton Abbot: David and Charles, 1969).
T. R. Nicholson, *The Birth of the British Motor Car, 1769–1897* (London: Palgrave Macmillan, 3 vols, 1982).
T. R. Nicholson, *Wheels on the Road: Maps of Britain for the Cyclist and Motorist, 1870–1940* (Norwich: Geo Books, 1983).
St John C. Nixon, 'Myra Edge – Appreciations', *Veteran Car Club Gazette*, 8 (Autumn 1969), p. 432.
St John C. Nixon, *The Antique Automobile* (London: Cassell, 1956).
St John Nixon, 'H. J. Lawson: The would-be dictator of the British motor industry', *Veteran and Vintage*, January 1957, pp. 167–8.
St John Nixon, *Romance Amongst Cars* (London: G. T. Foulis, 1937).
St John Nixon, *The Story of the S.M.M.T., 1902–1952* (London: SMMT, 1952).

Richard Noakes, 'Representing "A century of inventions": Nineteenth-century technology and Victorian *Punch*', in Louise Henson et al (eds), *Culture and Science in the Nineteenth-Century Media* (Ashgate: Aldershot, 2004), pp. 151–63.

D. Noble (ed.), *Royal Automobile Club: The Jubilee Book of the Royal Automobile Club, 1897–1947. The Record of a Historic 50 Years, etc.* (London: RAC Publishing, 1947).

Max Nordau, *Degeneration* (New York: D. Appleton & Co., 1895).

Peter D. Norton, 'Street rivals: Jaywalking and the invention of the Motor Age Street', *Technology and Culture*, 48:2 (2007), pp. 331–59.

Peter D. Norton, *Fighting Traffic: The Dawn of the Motor Age in the American City* (Cambridge, MA: MIT Press, 2008).

Sean O'Connell, *The Car in British Society: Class, Gender and Motoring, 1896–1939* (Manchester: Manchester University Press, 1998).

Nicholas Oddy, 'This hill is dangerous', *Technology and Culture*, 56:2 (2015), pp. 335–69.

Nicholas Oddy, 'The Anchor Hotel, Ripley: An analysis of the cyclists' books, 1881–1895', *Cycle History*, 13 (2002), pp. 108–13.

Nicholas Oddy, 'Cycling's Dark Age? The period 1900–1920 in cycling history', *Cycle History*, 15 (2004), pp. 79–86.

Ruth Oldenziel and M. Hård, *Consumers, Tinkerers, Rebels: The People Who Shaped Europe* (London: Palgrave Macmillan, 2013).

Samantha-Jayne Oldfield, 'Running pedestrianism in Victorian Manchester', *Sport in History*, 34:2 (2014), pp. 223–48.

George Oliver, *Cars and Coachbuilding: One Hundred Years of Road Vehicle Development* (London: Sotheby Parke Bernet, 1981).

Steven Parissien, *The Life of the Automobile: A New History of the Motor Car* (London: Atlantic Books, 2013).

P. L. Payne, 'The emergence of the large-scale company in Great Britain, 1870–1914', *Economic History Review*, 2nd ser., 20 (1967), pp. 519–42.

E. A. Penn, 'The wheel club', *The Boneshaker*, 114 (1987), pp. 4–5.

Emmeline Pethick-Lawrence, *My Part in a Changing World* (London: Victor Gollancz, 1938).

Frederick Pethick-Lawrence, *Fate Has Been Kind* (London: Hutchinson & Company Limited, 1943).

William Plowden, *The Motor Car and Politics in Britain, 1896–1970* (London: Harmondsworth, 1971).

D. Porter, 'Entrepreneurship', in S. W. Pope and J. Nauright (eds) *Routledge Companion to Sports History* (Abingdon: Routledge, 2010), pp. 197–215.

Christopher Thomas Potter, 'An exploration of social and cultural aspects of motorcycling during the interwar period' (unpub. PhD dissertation, Northumbria University, 2007).

R. G. G. Price, *A History of Punch* (Collins: London, 1957).

Carlton Reid, *Roads Were Not Built for Cars: How Cyclists Were the First to Push for Good Roads and Became the Pioneers of Motoring* (Newcastle on Tyne: Front Page Creations, 2014).

Kenneth Richardson, *The British Motor Industry, 1896–1939: A Social and Economic History* (London: Macmillan, 1977).

Jane Ridley and Clayre Percy, *Letters of Arthur Balfour and Mary Elcho, 1885–1917* (London: Hamish Hamilton, 1992).

Martin Riley, Bruce Lilleker and Neil Tuckett, *The English Model T Ford: A Century of the Model T in Britain* (Keighley: Model T Ford Register of Great Britain, 2008).

Andrew Ritchie, *Early Bicycles and the Quest for Speed: A History, 1868–1903* (Jefferson, NC: McFarland, 2nd ed., 2018).

Andrew Ritchie, *King of the Road: An Illustrated History of Cycling* (London: Wildwood House, 1975).

Derek Roberts, 'A short history of the Fellowship of Old Time Cyclists', *The Boneshaker*, 198 (2015), pp. 50–1.

L. T. C. Rolt, *The Horseless Carriage* (London: Constable, 1950).

W. D. Rubinstein, *Capitalism, Culture and Decline in Britain, 1750–1900* (London: Routledge, 1990).

Wolfgang Sachs (trans. Don Reneau), *For Love of the Automobile: Looking Back into the History of Our Desires* (Berkeley and Los Angeles, CA: University of California Press, 1992).

S. B. Saul, 'The motor industry in Britain to 1914', *Business History*, 5 (1962), pp. 22–44.

Virginia Scharff, 'Gender, electricity and automobility', in M. Wachs and M. Crawford (eds), *The Car and the City: The Automobile, the Built Environment, and Daily Urban Life* (Chicago: University of Michigan Press, 1992).

Virginia Scharff, *Taking the Wheel: Women and the Coming of the Motor Age* (New York: The Free Press, 1991).

Michael Brian Schiffer with Tamara C. Butts and Kimberley K. Grimm, *Taking charge: The electric vehicle in America* (Washington, DC: Smithsonian Institution Press, 1994).

Wolfgang Schivelbusch, *The Railway Journey: The Industrialization of Time and Space in the Nineteenth Century* (Berkeley, CA: University of California Press, 1986).

Barbara Schmucki, 'Against "the eviction of the pedestrian": The Pedestrians' Association and walking practices in urban Britain after World War II', *Radical History Review*, 114 (2012), pp. 113–38.

Jack Simmons, *The Victorian Railway* (London: Thames and Hudson, 1991).

Paul Smethurst, *The Bicycle: Towards a Global History* (London: Palgrave Macmillan, 2014).

E. J. Southcott (ed.), *The First Fifty Years of the Catford Cycling Club* (London: G. T. Foulis, 1939).

Ivan Sparkes, *An Illustrated History of Coaches and Coachbuilding: Stagecoaches and Carriages* (Bourne End: Spurbooks, 1975).

Susie L. Steinbach, *Understanding the Victorians: Politics, Culture and Society in Nineteenth-Century Britain* (London: Routledge, 2nd ed., 2017).

Ernest Stenson-Cooke, *This Motoring: Being the Romantic Story of the Automobile Association* (London: Automobile Association, 1931).

J. Steward, '"How and where to go": The role of travel journalism in Britain and the evolution of foreign tourism, 1840–1914', in J. K. Walton (ed.), *Histories of Tourism: Representation, Identity and Conflict* (Clevedon: Channel View, 2005), pp. 39–54.

J. Stobart and I. Van Damme, 'Introduction', in J. Stobart and I. Van Damme (eds), *Modernity and the Second-Hand Trade: European Consumption Cultures and Practices, 1700–1900* (London: Palgrave Macmillan, 2010).

J. A. Tarr and C. McShane, 'The decline of the urban horse in American cities', *Journal of Transport History*, 24:2 (2003), pp. 177–98.

J. A. Tarr and C. McShane, 'The horse as an urban technology', *Journal of Urban Technology*, 15:1 (2008), pp. 5–17.

Mikuláš Teich and Roy Porter (eds), *Fin de Siècle and Its Legacy* (Cambridge: Cambridge University Press, 1990).

Alun Thomas, 'The leviathan of the turf: R. H. Fry and the Grecian Villa', *Norwood Review*, Autumn 2017, pp. 11–18.

F. M. L. Thompson, 'Nineteenth-century horse sense', *Economic History Review*, 29:1 (1976), pp. 60–81.

J. Lee Thompson, *Politicians, Lord Northcliffe and the Press and the Great War, 1914–1919: Propaganda* (Kent, OH: Kent State University Press, 1999).

David Thoms, 'Motor car ownership in twentieth-century Britain: A matter of convenience or a marque of status?', in David Thoms, Len Holden and Tim Claydon (eds), *The Motor Car and Popular Culture in the Twentieth Century* (Aldershot: Ashgate, 1998), pp. 41–9.

Peter Thorold, *The Motoring Age: The Automobile and Britain, 1896–1939* (London: Profile, 2003).

G. A. Tobin, 'The bicycle boom of the 1890s: The development of private transportation and the birth of the modern tourist', *Journal of Popular Culture*, 7:4 (1974), pp. 838–49.

E. S. Tompkins, *Speed Camera: The Amateur Photography of Motor Racing* (London: G. T. Foulis, 1946).

John Tosh, 'Masculinities in an industrialising society: Britain, 1800–1914', *Journal of British Studies*, 44 (2005), pp. 330–42.

John Tosh, *Manliness and Masculinities in Nineteenth-century Britain: Essays on Gender, Family and Empire* (Harlow: Pearson, 2005).

Hugh Tracey, *Father's First Car* (London: Routledge & Kegan Paul, 1966).

Frank Trentmann, *Empire of Things: How We Became a World of Consumers, from the Fifteenth Century to the Twenty-first* (London: Penguin Classic, 2017).

Paul Tritton, *John Montagu of Beaulieu, 1866–1929: Motoring Pioneer and Prophet* (London: Golden Eagle/George Hart, 1985).

Ralph Turvey, 'Horse traction in Victorian London', *Journal of Transport History*, 26:2 (2005), pp. 38–59.

R. Turvey, 'Street mud, dust and noise', *London Journal*, 21:2 (1996), pp 131–48.

Alex Twitchen, 'The influence of state formation on the early development of motor racing', in Eric Dunning, Dominic Malcolm and Ivan Waddington (eds), *Sports Histories: Figurational Studies in the Development of Modern Sport* (London and New York: Routledge, 2004).

Michael Ulrich (ed., Thomas Ulrich), *Paris-Madrid: das grosste Rennen aller Zeitend* (Münster: Monsenstein and Vannerdat, 2015).

John Urry, 'The "system" of automobility', *Theory, Culture and Society*, 21:4–5 (2004), pp. 25–39.

Wray Vamplew, 'Sport, industry and industrial sport in Britain before 1914: Review and revision', *Sport in Society*, 19:3 (2016), pp. 340–55.

David Venables, *Racing Colours: British Racing Green: Drivers, Cars and Triumphs of British Motor Racing* (Shepperton: Ian Allen, 2008).

David Venables, *Napier: The First to Wear the Green* (London: G. T. Foulis, 1998).

Alan Vessey, *By Precision into Power: A Bicentennial Record of D. Napier and Son* (Stroud: Tempus, 2007).

Margaret Walsh, 'Gendering transport history: Retrospect and prospect', *Journal of Transport History*, 23:1 (2002), pp. 1–8.

Martin Walter, '"The song they sing is the song of the road": Motoring and the semantics of space in early-twentieth-century British travel writing', *Transfers*, 5:2 (2015), pp. 23–41.

J. K. Walton, 'The origins of the modern package tour? British motor-coach tours in Europe, 1930–1970', *Journal of Transport History*, 32:2 (2011), pp. 145–63.

Michael E. Ware, 'The 1896 Emancipation Run', in Malcolm Jeal (ed.), *London to Brighton Run (Centenary)* (Crawley: Consortium, 1996), pp. 13–18.

John Whiteley, 'The beginning of the motor age, 1880–1940', in Alan Crosby (ed.), *Leading the Way: A History of Lancashire's Roads* (Preston: Lancashire County Books, 1998), pp. 183–239.

Martin Wiener, *English Culture and the Decline of the Industrial Spirit, 1850–1980* (Cambridge: Cambridge University Press, 2nd ed., 2004).

Jean Williams, *A Contemporary History of Women's Sport, Part One: Sporting Women, 1850–1960* (New York and London: Routledge, 2014).

Alice Williamson, *The Inky Way* (London: Chapman and Hall, 1931).

Charles Wilson and William Reader, *Men and Machines: A History of D. Napier and Son, Engineers Ltd., 1808–1958* (London: Weidenfeld & Nicolson, 1958).

Reg Winstone, 'Grand Prix cyclecar', *The Automobile*, August 2019, pp. 22–30.

Sarah Wintle, 'Horses, bikes and automobiles: New Woman on the move', in Angelique Richardson and Chris Willis (eds), *The New Woman in Fiction and Fact: Fin de Siècle Feminisms* (London: Palgrave, 2002), pp. 66–78.

Abigail Woods, 'Rethinking the history of modern agriculture: British pig production, c. 1910–65', *Twentieth-Century British History*, 23:2 (2012), pp. 165–91.

Michael Worthington-Williams, 'Henry Dawson and the Canterbury', *The Automobile*, October 2007, pp. 48–51.

Michael Worthington-Williams, 'From "New Motoring" to the microcar', *Aspects of Motoring History*, 13 (2017), pp. 5–20.

Michael Worthington-Williams, *From Cyclecar to Microcar* (London: Dalton Watson, 1981).

Julie Wosk, *Women and the Machine: Representations from the Spinning Wheel to the Electronic Age* (Baltimore, MD: Johns Hopkins University Press, 2001).

Horace Wyatt, *Common Commodities and Industries: The Motor Industry* (London: Sir Isaac Pitman, n.d. [c. 1920]).

Appendix

Biography

Joseph ('Johnny') Henry Adams (*c.* 1863–1932), racing cyclist, promoter of Germain motor cars
Daniel Albone (1860–1906), cyclist, promoter of motor-tractors
Albert Argent Archer (1860–1932), photographer
Henry Thwaite Arnott (b. *c.* 1854), club cyclist, promoter of Princeps motor-cycles and cars
Herbert Austin (1866–1941), of the Wolseley Motor Co until 1905, thereafter manufacturer
Henry Somerset, ninth Duke of Beaufort (1824–99), editor, associated with the Badminton series
Sydney Dawson Begbie (*c.* 1865–1947), cyclist; manager, Century cars
F. T. Bidlake (1867–1933), racing cyclist, journalist
Gerald Biss (1876–1922), motoring journalist
Charles William Brown (b. *c.* 1866), journalist, cyclist, early motorist
Roland Browne, motorist, horseman, and **Mrs Roland Browne**, motorist
T. B. Browne (1873–1965), motorist, and **Mrs T. B. Browne**, motorist
Henry Burford (1867–1943), motor sales engineer
Alfred Burgess, secretary, MMC cars
Robert Burns, manager, Swift cycles and cars
Frank Hedges Butler (1855–1928), pioneer motorist, balloonist, big-game hunter
Captain Kenneth Campbell (b. 1863), motorist
Ernest Harcourt Coles (1865–1925), 'trick' driver with Benz, engineer
Mrs C. C. Cooke, contributor, *Motorcycling* magazine
Major George Cornwallis-West (1871–1951), soldier, sportsman
(James) Sidney Critchley (1865–1944), mechanical engineer
'Cyclomot' [George F. Sharp] (1868–1928), journalist; editor, *The Motor Boat* (1904); editor, *The Commercial Motor* (1917–28)
Edmund Dangerfield (1864–1938), founder editor of *Cycling* and *The Motor*
Harriet Dibble (d. 1887), landlady, the Anchor Inn, Ripley
Harvey du Cros (1846–1918), cyclist, businessman, Conservative MP, associated with Dunlop
Arthur Du Cros (1871–1955), cyclist, businessman, Conservative MP
Camille du Gast (1868–1942), racing driver
Helene Dutrieux (1877–1961), cyclist, racing driver, aviator
Herbert Osbaldeston Duncan (1862–1945), cycling promoter, associated with De Dion, Paris-based
Selwyn Francis Edge (1868–1940), racing cyclist, racing driver, motor salesman, associated with Napier cars
Hubert Egerton (1875–1950), racing cyclist, motorist

Maurice Egerton (1874–1958), motorist, aviator
Thomas Robert Barnewall Elliot (*c.* 1871–1949), first motorist in Scotland
Henri Farman (1874–1958), cyclist, racing driver, aviator
Alexander Bell Filson Young (1876–1938), writer, contributor to the *Daily Mail*; wrote *The Happy Motorist* and *The Complete Motorist*
Lawrence Fletcher (b. *c.* 1861), early Cyclists' Touring Club member, active in the Anfield Bicycle Club
Prince Francis of Teck (1870–1910), president of the Automobile Club
Charles Lincoln Freeston (1865–1942), travel writer
Charles Friswell (1871–1926), importer of Peugeot cars, associated with the Automobile Mutual Protection Association Ltd
Richard Henry Fry (1836–1902), president, Anerley Bicycle Club; businessman
F. R. Goodwin, racing cyclist, manager at Star cycles and cars
T.W. Grace, solicitor, a 'noted runner and cyclist', 'leading light' of the Manchester Automobile Club
Claude Grahame-White (1879–1959, brother of Montague), motor engineer, aviator
Montague Grahame-White (1877–1961), playboy, motorist
Samuel Wilson Greasley (1867–1926), long-distance swimmer
John Haggerty (1862–1939), long-distance swimmer
Alfred Harmsworth (1865–1922, later **Lord Northcliffe**), journalist; editor, *Daily Mail*
Henry Hewetson (1852–1930), pioneer motorist, agent for Benz
Muriel Hind (1882–1956), motor-cyclist
James John Hissey (1847–1921), travel writer
Montague Holbein (1861–1944), racing cyclist, long-distance swimmer
Ernest Terah Hooley (1859–1947), motoring financier
George W. Houk, promoter of American steam cars
Ernest Instone (1872–1932), motorist, associated with Automobile Club and SMMT
Charles Jarrott (1877–1944), pioneer motorist, racing driver
W. Rees Jeffreys (1871–1954) secretary, Roads Improvement Association
Lady Mary Jeune (1845–1931), motorist, writer
Claude Johnson (1864–1926), Secretary, Automobile Club
Basil Joy (1870–1940), Technical Secretary, Automobile Club
Mary Kennard (*c.* 1844–1936), journalist, fiction writer, motorist prior to 1900
Sir Vincent Kennett-Barrington (1844–1903), balloonist
John Adolphus Koosen (*c.* 1860–1913), pioneer motorist
Harry Lawson (1852–1925), motoring financier, founder of the Motor Car Club
William Letts (1873–1957), motoring agent, Locomobile
Dorothy Levitt (1882–1922), racing driver
Major F. Lindsay Lloyd (1866–1940), clerk of the course, Brooklands
'Owen John' (**Owen John Llewellyn**) (1870–1943), motoring writer
Major Charles George Matson (1859–1914), writer, journalist, wrote *The Modest Man's Motor* (1903)
Mark Mayhew (1871–1944), racing driver, businessman, soldier
George Pilkington Mills (1867–1945), racing cyclist, associated with Anfield Bicycle Club
Lord Montagu of Beaulieu (1866–1929), editor, *The Car Illustrated*
Henry Vollam Morton (1892–1979), travel writer and social commentator
Walter Munn (b. 1863), cyclist; secretary, De Dion Bouton
Montague Napier (1870–1931), cyclist, manufacturer of Napier cars
Alice Hilda Neville, motoring businesswoman

Sheila O'Neil, London taxi driver
Frank Patterson (1871–1952), motoring illustrator
Louis Paulhan (1883–1963), aviator, first to fly London to Manchester
Max Pemberton (1863–1950), motoring writer, wrote *The Amateur Motorist*
Edward J. Pennington (1858–1911), motoring financier, motor manufacturer in the 1890s
Percival Perry (1878–1956), motor agent for Ford
Toby Rawlinson (1867–1934), motor entrepreneur, esp. with Darracq
Charles Rolls (1877–1910), motor salesman, motorist, aviator
David Salomons (1851–1925), scientist, motorist, associated with Self-Propelled Traffic Association
Eugen Sandow (1867–1925), strongman, showman
Alberto Santos-Dumont (1873–1932), balloonist
James Selby (1844–88), coachman, set the London-Brighton-London for a coach in 1888
Eleanor Rose Sharp (1872–1914), motorist, m. S. F. Edge
Frank Shorland (1871–1929), racing cyclist, motoring entrepreneur
Edward Shrapnell-Smith (1875–1952), manager of the Road Carrying Company
Frederick Simms (1863–1944), held UK patents for Daimler, founder Automobile Club
Paris Singer (1867–1932), businessman, City and Suburban Electric Carriage Company, benefactor of the Automobile Club
George H. Smith (1862–1946), cyclist, motoring entrepreneur, kept diary of Anerley Bicycle Club
Percival Spencer (1864–1913), balloonist
Arthur Stanley (1869–1947), chairman of the Automobile Club, Conservative MP
John William Stocks (1871–1933), racing cyclist, manager, De Dion Bouton
Henry Sturmey (1857–1930), journalist; editor, *The Cyclist*; editor, *The Autocar*
J.B. Thistlethwaite, an 'all-round sportsman'; secretary, Manchester Automobile Club
Charles McRobie Turrell (1875–1923), pioneer motorist, right-hand man to Harry Lawson
James Edmund Vincent (1857–1909), travel writer
George Henry Wait (b. 1867), cyclist; manager, Clyde cycles and cars
Frank Frederick Wellington (1868–1917), pioneer motorist, engineer
Alice Williamson (1858–1933), motorist, motoring novelist, with husband C. N.
W. Williamson (b. 1873), managing director, Rex motor-cycles and cars
Joseph van Hooydonk (1866–1936), cycle and motor-cycle manufacturer
Helene van Zuylen (1863–1947), racing driver, competed in the 1898 Paris-Amsterdam-Paris run

Index

alcohol   148, 167
Arts and Crafts (movement)   16

Bartholomew (maps)   131
BBC   131, 177
British Covenant for Ulster   60 n.108

camping   12, 122, 134, 135, 136, 140, 152, 159
caravanning   134, 134 n.85, 135
charabanc   136, 137, 140
Conservative (party, MP)   45, 60, 61, 63, 78, 148

Eton (college)   44

firearms   26, 118, 152

L. B. and S. C. Railway   14
Lever Bros   1
Liberal (party, MP, candidate)   25, 61, 62, 78, 153, 159
Locomotives Act (1865)   15
Locomotives on Highways Act (1896) ('Emancipation Act')   13, 15, 123, 137

Maxim (machine gun)   152
Mother Shipman's (soap)   1

Motor Car Act (1903)   137
Murray (guides)   131

Ordnance Survey   131

Parkyns-v-Priest   16
Pear's (soap)   1

'rational' (dress)   4, 16, 38, 59, 63, 173
Royal Commission   58

Sandhurst (college)   44
shale oil   148
South African War   53
speed trapping   11, 12, 36, 121, 122, 131, 137, 138, 139, 174
Stepney (wheel)   114, 120
strike, Liverpool General Transport   152
Sunday/Sabbath   4, 33, 40, 121
Sunlight (soap)   1

tank   152
Transvaal War Fund   71
'trimo'   155, 155 n.71, 156

Warwickshire hounds   26

'yokel'   28, 135, 139, 143
Yost (typewriters)   58

Clubs, organisations

AA, *see* Automobile Association
ABC, *see* Anerley Bicycle Club
ACGBI, *see* Automobile Club of Great Britain and Ireland
Aero Club   144, 145
Aeronautical Society   145

Anerley Bicycle Club   22, 26, 27, 31, 33, 46, 47, 48, 49, 58, 59, 60, 62, 97, 113, 115, 116, 126, 128, 177
  'Old Boys Section'   49
Anfield Bicycle Club   113, 126

Association for the Promotion of
    Flight   145
Association of Cycle Campers   134
Auto-Cycle Club   84, 157, 160, 161
Automobile Association   29, 36, 84,
    113, 138
Automobile Club of Great Britain
    and Ireland   12, 25, 29, 30 n.110,
    37, 48, 57, 65, 72, 78, 87, 112,
    113, 114, 115, 116, 118, 120,
    121, 126, 133, 147, 157, 162,
    171, 176
  Campbell's paper   103, 104
  Easter tour   76
  formation   112 n.116, 151 n.40
  Jarrott's membership   50
  'maker's amateurs', attitude   35, 44, 45
  membership, debates over   34, 35, 36
  The Motor, spat   82, 83, 84, 85, 86
  Thousand Mile Trial,
    organisation   66, 67, 68, 70, 71,
    73, 74
Automobile Mutual Association   36
Automobile Mutual Protection
    Association Ltd   17

Bath Road Cycling Club   45, 46, 91 n.9
'Black Anfielders', *see* Anfield Bicycle Club
Blackburn Cycling Club   31
Board of Agriculture   39
British Air Services   146

Catford Cycling Club   31, 32, 45, 46,
    49, 178
Circle of Nineteenth-Century
    Motorists   111, 176
Coaching Club   24, 31
Committee of Imperial Defence   146
Connaught Club   33
CTC, *see* Cyclists' Touring Club
Cyclecar Club   160
Cyclists' Touring Club   7, 121, 126

East India Company   44

Fellowship of Old Time Cyclists   112,
    177, *178*
Four in Hand Club   24, 31

Herefordshire Automobile Club   72

'K9' Society   65

[Launceston] City Cycling
    Club   35 n.131
Liverpool Self-Propelled Transport
    Association   77, 77 n.47, 151

MAC, *see* Manchester Automobile
    Club
Malvern Cycling Club   32
Manchester Automobile Club   72, 75, 78,
    79, 87, 113, 115
Manchester Motor Club   78, 79, 87,
    113, 134
Marine Motor Association   147
Ministry of Munitions   153
MMC, *see* Manchester Motor Club
Motor Car Club   13, 14, 55, 56, 133
Motor Cycle Club   31, 49
Motor Cycling Club   39
Motor Trades Association   48
Motor Union   36, 84, 113

National Cyclists' Union   35
North Road Cycling Club   33

People's Refreshment House
    Association   116
Polytechnic Cycling Club   98
Preston Cycling Club   31
Primrose League   60 n.108
Pure Food and Health Society   57

RAC, *see* Automobile Club of Great
    Britain and Ireland
Road Board   162
Road Club   31
Rossendale Bicycle and Tricycle Club   31
Royal Flying Corps   39
Royal Navy   146
Royal Norfolk Automobile Club   154

Scottish Automobile Club   75
Scottish Motor Trade
    Association   43 n.10
Self-Propelled Transport
    Association   151

SMMT, *see* Society of Motor
    Manufacturers and Traders
Society of Motor Manufacturers and
    Traders   36, 46, 48, 49, 58, 61, 82,
    82 n.76, 85, 85 n.88, 87, 145, 151 n.40
SPTA, *see* Self-Propelled Transport
    Association
Sydney Bicycle Club   35

25th County of London Cyclist
    Regiment   152

Union Jack Club   62
Union Jack Industries League   62

Vale of Lune Cycling Club   32
Veteran Car Club   55 n.66
Veteran-Cycle Club   9

Watford Cycle Club   49
Wheel Club   33, 34 n.128
Women's Unionist and Tariff Reform
    Association   60 n.108

Events, awards, races, circuits

Battle of Flowers   59 n.103, 117
British International Cup for Motor-
    Boats   147
Brooklands (circuit)   38, 53

Challenge Cup   145
Cromer   76
Crystal Palace Motor Show   82, 91, 147
Cuca Cocoa Challenge Cup   47

Easter Tour (1889)   127
Easter Tour (1898)   76, 121
Edge's 24-hour non-stop driving
    record   53, 143
'Edge's Run'   33, 49
'Emancipation Run'   13, 14, 16, 17, 18,
    24, 67, 68, 70, 89, 94, 163, 167,
    170, 176

5,000 Mile Trial   75, 105
French Charity Flowers show   116

German Exhibition   20
Gordon Bennett Cup   8, 19, 36, 53, 54,
    55, 57, 65, 74, 76, 77, 174
Gunton Park   76

International Cycling History
    Conference   9, 9 n.42

Light Car Trial   65, 72, 78, 118
London-Brighton-London   47, 48
London-Liverpool-London   118
'London to Brighton Run', *see*
    'Emancipation Run'
London-York   47

M25 (road)   163
Mediterranean Cup   147
Michael Argetsinger Symposium   42 n.9
Midland Cycle and Motor Show   76
Motor Show   120

Olympia Motor Show   56, 147, 164 n.116

Paris-Amsterdam-Paris   60
Paris-Bordeaux-Paris   74
'Paris in London' fête   117 n.140
Paris-Innsbruck   54
Paris-Madrid   74, 75, 167
Paris-Marseille (1896)   51
Paris Motor Show   18, 154
Paris-Vienna   54
Reliability Trials for Motor-cars   117
Richmond Show   67

Scottish Trial   75 n.39
S. F. Edge Novice's Cup   49
S. F. Edge Old Boy's Cup   49

S. F. Edge Trophy   49
Six Days' Trial   160
Snake Hill   79
Stanley Cycle Club dinner   48 n.30
Stanley Cycle Club Show   15, 119

The Motor light car trial   83, 84, 85, 86

Thousand Mile Trial   5, 11, 25, 28, 35, 51, 66, 67, 68, 69, 70, 71, 72, 73, 74, 75, 77, 78, 111, 113, 126, 130, 133, 149, 164, 165, 171, 176
'Tour to Brighton', *see* 'Emancipation Run'

Voiturette Trial   97 n.33

Publications, publishers

The Aero   145
Aero Manual   145
Agricultural Gazette   58 n.98
Anticipations   162
Aspects of Motoring History   9 n.41
The Autocar   2, 12, 18 n.30, 26, 29, 34, 44, 53, 57, 66, 69, 71, 73, 75, 76, 77, 78, 80, 81, 82, 85, 85 n.88, 97, 101, 105, 109, 121, 122, 138, 157, 158, 161 n.101, 162, 167, 170, 173
Automotor and Horseless Vehicle Journal   2, 18 n.30, 19, 68, 71, 145

Badminton Library of Sports and Pastimes   23, 57, 103, 148, 156
Badminton Magazine   18 n.30, 102
Bicycling News (*Bicycling News and Motor Car Chronicle*)   10, 84, 126, 133
The Boneshaker   9 n.42
Bystander   18 n.30

The Car Illustrated   28, 29, 30, 35, 38, 59, 117 n.140, 123, 136, 148, 152, 157, 163, 174, 177
The Chauffeur   77
The Clarion   38, 69
Commercial Motor   144, 151
The Complete Motorist   131
Cornwall and a Light Car   131
Country Life   58 n.98, 117, 135 n.95
Cycle and Motor Trades Review   34
The Cyclecar   108, 116, 133, 135, 144, 158, 159, 160, 167, 178
Cycling   10, 33, 44, 45, 48, 100, 101
The Cyclist   7, 44

Daily Mail   10, 17, 18, 18 n.30, 24, 25, 30, 61, 62, 68, 76, 82, 95, 103, 107, 131, 146, 173
Daily Sketch   60
Daily Telegraph   147
Degeneration   16

The Engineer   61

F. King and Co   145
The Flag   62
Flight   144

The Gentlewoman   117 n.140
Girl's Own Paper   102
The Graphic   117 n.140

The Happy Motorist   131
Harmsworth London Magazine   3
Hearth and Home   70

Iliffe   85 n.88
Illustrated London News   1, 156, 177, 177 n.23
Implement and Machinery Review   58 n.98

Journal (ACGBI)   35, 83

King Solomon's Mines   125 n.24

Ladies' Pictorial   28
The Lady Cyclist   116
The Light Car   158
The Lightcar and Cyclecar   138, 159 n.84
The Lightning Conductor   41, 123

Manchester Chronicle   134
Modest Man's Motor   102, 103
Monthly Gazette (CTC)   121
The Motor, see *Motorcycling and Motoring*
The Motor Boat   144, 147
Motor Car Journal   11, 28, 68, 73, 149
The Motor Cycle   85 n.88
Motorcycling   29, 37, 39, 60, 90, 100, 106, 108, 117, 127, 133, 143, 144, 152
Motorcycling and Motoring   10, 11, 19, 25, 28, 29, 31, 33, 38, 44, 45, 66, 78, 79, 80, 85 n.88, 88, 91, 92, 94, 95, 96, 97, 98, 108, 110, 114, 116, 118, 120, 134, 136, 137, 138, 147, 149, 152, 154, 155, 156, 157, 161, 163, 169, 175
   circulation   74 n.37, 101 n.54
   Light Car Trial   82, 83, 84, 85, 86
   'modest motoring'   83, 84, 87, 93, 101, 102, 103, 104, 105, 106, 109, 112, 158, 160, 168, 170, 178
   spin-off magazines   87, 100, 144, 151, 158, 178
   starting up   100, 101, 102

Motoring Illustrated   19, 59, 105, 109, 110, 139, 144, 148, 157, 166
The Motor Maniac   25, 41, 59
Motors and Motor Driving   148
Motor Trader   46

Penny Illustrated Paper   57, 164
Punch   2, 3, 10, 11, 14, 19, 22, 25, 26, 38, 39, 43, 52, 60, 75, 108, 113, 116, 135, 137, 139, 143, 144, 146, 147, 150, 153, 164, 166, 170, 173

The Scotsman   43
Society   70
The Sphere   139
The Strand   3, 5, 20, 132, 143, 150, 169

The Tatler   38
Temple Press   10, 99, 144, 145
The Times   14, 19, 20, 24, 26, 57, 61, 71, 147, 172, 176, 177

Whitaker's Almanac   104 n.67
The Woman and her Car   60, 117

Places

Algiers   147
Alps   121
Alton   22
Arundel   127
Ascot   124
Australia   56, 57, 58
'Automobile Palace'   55

Barnet   84
Belgium   68
Belvedere (Kent)   109
Ben Nevis   164
Bettws-y-Coed   79
Bicester   145
Birkhill   73
Birmingham   68, 72, 82, 109, 157 n.81
Black Country   135
Bowden (Ches)   104

Brighton   13, 14, 15, 24, 28, 39, 68, 76, 133, 136, 145, 162
   Metropole Hotel   13, 176
Brighton Road   6, 36, 126, 138, 162, 163, 172
Bristol   68
Bromley   84
Buxton   69

Caithness   153
Canterbury   51, 139
Carlisle   68, 73, 138
Cheltenham   73
Cheshire   79, 93, 104, 106, 107, 112, 112 n.117, 118
   Tatton Hall   44, 145
Clitheroe   157 n.81
Cobham
   White Lion inn   29 n.106

Connaught [Connacht]   153
Cornwall   166
Coventry   27, 43, 58, 81
Crewe   112 n.117
Cricketers (pub)   127
Crystal Palace   47, 57, 70, 74, 78, 84, 98

Darlington   51, 69
Dashwood Hill   76
Deauville   146
Derbyshire   79
Devon   93, 166
Dieppe   121, 128
Ditton
   Angel inn   29 n.106
Dublin   79

East Anglia   131
Edinburgh   43, 50, 66, 68, 73, 75, 76, 119,
   149, 164, 166
England   5, 14, 15, 18, 19, 25, 28, 55, 79,
   81, 95, 121, 124, 131, 140, 176
Epping   29
   Ye Olde Thatched House   29
Esher
   The Bear inn   29 n.106

Folkestone   163
France/French   3, 13, 14, 18, 19, 68, 81,
   97, 101, 116, 121, 128, 131, 143,
   146, 157

Germany/Germans   3, 13, 16, 19, 21, 50,
   61, 92, 130, 151 n.40, 152 n.51
Gladsmuir   149
Glasgow   58
Glossop   79, 134

Hampshire   138
Hawarden Castle   149
Hayward's Heath   162, 162 n.105
Hereford   65, 66, 68, 72, 86
Holyhead   79
Ireland   53, 153
Islington   128
John O'Groat's   1, 7, 33 n.126, 121, 128
Kendal   68
Kensington   148
Kent   5, 21, 28, 151 n.44

Lake District   5, 138, 160, 166
Lancashire   136
Land's End   1, 33 n.126, 121, 128
Lansing (Michigan)   164 n.117
Leamington   80
Leeds   68, 73, 130, 171
Leicester   28
Lewes   152
Lincolnshire   93
Liverpool   126, 152
London   2, 20, 29, 43, 45, 46, 58, 59, 66,
   70, 71, 73, 75, 80, 81, 84, 89, 95, 108,
   119, 128, 129, 130, 139, 145, 146,
   149, 150, 162, 163, 166, 167, 170, 171
   Alhambra   164
   Belvedere House College   44
   Bond Street   56
   Bridge House Hotel (Southwark)   48
   Earls Court   117
   Embankment   151
   Frascati's Restaurant   31
   Great Portland Street   157 n.81
   Grosvenor Place   171
   Holborn Restaurant   48
   Holborn Viaduct   50, 55
   Horse Guards Avenue   171
   Hyde Park   20
   Imperial Institute   55, 56
   Inns of Court Hotel   33
   Midland Grand Hotel (St Pancras)   18
   New Burlington Street   55
   Old Kent Road   139
   Piccadilly (Circus)   20, 111 n.111
   Purley Corner   145
   Royal Exchange   23
   Tooting   157 n.81
   Trocadero Restaurant   111 n.111
   Upper Norwood   44
   West Norwood   57
   Whitehall Court   35, 55
Lowther Castle   43
Lymm   104

Maidstone   139
Manchester   29, 67, 68, 78, 79, 104, 113,
   130, 136, 146, 151, 161, 165, 167, 168
   Deansgate   23
   Grand Hotel   29
   Royal Botanical Gardens   70, 78

Margate   2, 89, 129, 139
Midlands   162
Monaco   146
Monte Carlo   119

Nantwich   78, 78 n.54
Newcastle   68, 149
New Zealand   80
Nîmes   3 n.17
Northallerton   51
'North Road'   128
Nottingham   61, 68

Offington   127

Paris   48, 50, 71, 89, 157, 157 n.81
Persia   134
Portsmouth   29, 95, 163
Purley   126
Putney   25

Ramsgate   139
Ranelagh   40, 145
Riddlesdown   127, 128
   Rose and Crown (pub)   128
Ripley   27, 29, 127
   Anchor Inn   29
Robin Hood (pub)   126

St Andrews   28
Salford   78, 104
Sandringham   117 n.144
Sandwich   139
Scotland   5, 43, 50, 128, 166
Selsey
   Crown inn   46 n.21

Sheffield   29, 68
South Africa   56
South America   57
Southsea   69
Spain   74
Surrey   5, 139
Sussex   5, 49
Switzerland   133

Taddington   73
Thames (river)   55, 127
Torquay   98
Toulon   147
Tranent   149

United Kingdom   1, 10, 11, 13, 14, 18, 45, 56, 58, 66, 68, 74, 97, 145, 153 n.55, 164 n.117
United States   42, 56, 61, 62, 132 n.76, 153
Upminster   139

Wales   5, 126, 131, 136, 170
Warwick   26
Welbeck Abbey   36, 76
West Wycombe   76
Wiltshire   2
Winchester   139
Wisley   29
   Hut Hotel   29
Worcestershire Beacon   164
Worthing   60

Yorkshire   78

Zambesi (river)   147

People

Adams, J. H.   45
Albone, Daniel   152
Alexander, Henry   164
Alexis and Dorrano ('adagio
   dancers')   111 n.111
Amini, Shima   18 n.26
Archer, John   78, 79

Argent Archer, Albert   67, 68, 70, 72, 78, 124, 126
Armstrong, A. C.   101, 144
Arnott, H. T.   46
Ashton, T. S.   42
Asquith, H. H.   146
Austin, Herbert   54, 76, 77, 112

Index

'Autolycus' 134
'Automan' 93, 107, 110

Bacon, N. G. 102
Baden-Powell, Robert 146
Balfour, Arthur 56, 61, 149, 153
Banfield, Frank 153
Bartleet, 'Sammy' 34, 34 n.128
Bateman, Arthur 15
Batt, Harry 24
Beaufort, Duke of 23, 23 n.61, 24
Beckett, H. R. 81
Begbie, S. D. 46
Bennett, Elizabeth 68
Bersey, Walter 43 n.10
Bidlake, F. T. 21, 68, 70, 165 n.122, 171, 172
Bierbaum, Otto 122
Biss, Gerald 38
Blew, W. C. A. 23
Bolster, John 105
Boutle, Ian 53
Brendon, Piers 34, 72, 176
Breward, Christopher 55
Bridgman, G. S. 98
Brooks 51, 59
Brown, C. W. 10, 10 n.50, 84
Browne, Mrs Roland 148
Browne, Mrs T. B. 148
Browne, T. B. 170, 172
'Bull, John' 1
Burford, Henry 37
Burgess, A. 46
Burgess, W. H. M. 22
Burgess-Wise, David 9 n.41
Burns, R. 46
Butt, Josh 22 n.53, 23
Buzard, James 124

Campbell, Capt. Kenneth 103, 104, 105
'Cecil-Lanstown, Reginald' 41
Chiozza Money, L. G. 159
Churchill, Winston 78
Clarsen, Georgine 59
Clausager, Anders Ditlev 9 n.41, 19 n.32
Clayton, Nick 177
Clifford Earp, W. T. 36
Coles, E. H. 73, 164, 166 n.135
Cooke, Mrs C. C. 60, 117

Cornell, Alfred 33
Cornwallis-West, Major George 50
Coulbert, Esme 123, 124, 132
Critchley, James Sidney 37, 43 n.10
Cusins, Muriel 97, 97 n.33
'Cyclomot' 25, 96, 103

Dangerfield, Edmund 44, 45, 85, 85 n.88, 91 n.9, 101, 144, 147, 154, 158, 159, 178
Davies, B. H. 155
Davis, Harry 34
Denning, Andrew 4, 122
Dibble, Harriet 29, 127
Du Cros, Arthur 55, 55 n.70, 61
Du Cros, bros 58
Du Cros, Harvey 35, 38, 61
Dudley, Countess of 37, 153
Dudley, Earl of 153
Duffy, Enda 5, 11, 122
Du Gast, Camille 60
Duncan, H. O. 17, 18, 19, 44, 48, 51, 54, 97, 154 n.68
Dutrieux, Hélène 59

Edge, Alexander 46, 57, 127
Edge, Arthur Cecil 76
Edge, Eleanor, *see* Sharp, Eleanor
Edge, Henry 57
Edge, Kelburne 35, 35 n.131, 46, 58
Edge, Myra, *see* Martin, Myra Caroline
Edge, Seaton 46
Edge, Selwyn Francis 7, 8, 11, 23 n.62, 26, 32, 36, 37, 40, 42, 45, 48, 51, 52, 52, 57 n.85, 63, 65, 68, 72, 75, 76, 82, 83, 84, 90, 92, 97, 98, 101, 111, 112, 116, 117, 118, 119, 128, 129, 131, 143, 148, 152, 152 n.53, 154, 168, 170, 178
   aviation, attitude 145, 146
   charity work 57
   cultivating image 53, 54, 55, 56, 143, 177
   cycling clubs 46, 47, 176
   cycling records 47
   Director of Agricultural Machinery 39, 153
   family life 44, 46, 57, 58
   Gordon Bennett Cup 54, 174

guaranteeing cars   56, 176
horse traction, attitude   164, 165, 172
meeting Jarrott   50
membership of ACGBI   35, 176
motor boating   147
patents   51
'Pellin Sedge'   59
pig breeding   31, 58
politics   60, 61, 62
presidencies   31, 33, 49, 61, 62, 120, 177
sales methods   43, 44
Sandow's exercises   53, 54
social fit   38, 39
sponsorship   49
travel abroad   56
violence   27, 28, 126
writing autobiography   54, 57
writing to the press   19, 20, 24, 34, 57, 95, 172
Edge, William   26
Egerton, Hubert   97, 155 n.68
Egerton, Maurice   44, 44 n.14, 145
Elcho, Lady Mary   56 n.76
Elliott, T. R. B.   50, 112 n.114
Ellis, Evelyn   151 n.40
Estensen, Corey   44 n.14

Farman, bros   71
Farman, Henri   37
Fawcett, E. Douglas   133
Filson Young, Alexander Bell   29, 42, 43, 105, 125, 130, 131, 132, 135
Finch-Hatton, Henry   13
'Fitztights, Maudie'   77
Fletcher, J.   28
Fletcher, Lawrence   126
Flower, Charles   174, 175
Ford, Henry   112
Francis, Prince, of Teck   78
Frankland-Payne-Gallwey, Ralph   26
Franz, Kathleen   80
Fraser   126
Freeston, C. L.   123, 131
Friswell, Charles   17, 46, 55
Fry, R. H.   49, 60, 61, 120
Fuller, Roger   *36*

Gelber, Steven M.   167
Genus, Audley   134

Georgano, G. N.   9 n.41
Gladstone, William   149
Goodwin, F. R.   46
Goodwin, Mr   1
Gordon, W. J.   20
Grace, T. W.   78
Grahame-White, Claude   146
Grahame-White, Montague   35, 43 n.10, 66, 96, 111, 125, 125 n.24, 146, 176
Granby, Marquis of   152 n.53
Greasley, Samuel Wilson   147
Green, Harry   34, 34 n.128
Grey, Charles   145
Groves, Walter   33, 101
Gurney, Goldsworthy   15, 16

Haggard, H. Rider   125
Haggerty, John   147
Hahn, H. Hazel   55, 80
Hales, David   36
'Hanks, Mr'   3
Harmsworth, Alfred   7, 10, 25, 35, 59, 61, 69, 82, 131, 146, 148, 153, 156, 167, 173
Harmsworth, R. Leicester   153
Harper, Martin   31
Harrison, A. E.   9, 17
Harrison, John   164 n.118
Hart-Davies, Ivan B.   106
'Harvey'   41
Hearne, R. P.   150
Hedges Butler, Frank   124, 125, 125 n.24, 144, 145
Hewetson, Henry   130, 169
Hewetson's (agent)   164
Hills, Dr   109
Hind, Muriel   60
Hissey, J. J.   5, 21, 28, 123, 125, 137
Hogan, B. H.   34, 34 n.128
Holbein, Montague   147
Hooley, Ernest Terah   17, 18, 55
Hope, John   107 n.85
Horsfield, F.   134
Houk, G. W.   46
Howe, Earl   61
Hughes, H. S.   22

Instone, Ernest   43 n.10, 45, 61
Ivory, Chris   134

Jarrott, Charles 2, 19, 29, 36, *36*, 37, 38, 39, 54, 56, *83*, 84, 90, 95, 97, 110, 117 n.140, 118, 125, 131, 138, 140, 148, 152, 152 n.53, 154
   'Charlemagne Parrott' 59
   looking backwards 15, 22, 26, 91, 119, 129
   Margate to London 89
   personality 45, 50
   prosecution 26
   violence 27, 27 n.87, 27 n.90
Jeal, Malcolm 9 n.41, 16, 58 n.94, 75
Jeffreys, W. Rees 162
Jenatzy, Camille 5, 143
'Jenks, Mrs Janet' 41, 59
Jeremiah, David 122, 123, 136, 164
Jeune, Lady Mary 38, 59
Johnson, Claude 31, 35, 44, 66, 67, 68, 103 n.63, 162
Jones, J. C. 166
Joy, Basil 85

Kennard, Mary 19, 25, 28, 41, *51*, 59, 133
Kennett-Barrington, Sir Vincent 145
Knox, C. Uchter 138
Koosen, J. A. 2, 130

Langridge, G. T. 162 n.105
Lansdowne, Lord 56
Larkin, Leonard 169, 170
Larkins, Miss J. 118
Law, Michael John 10, 136
Lawson, Harry 13, 14, 16, 17, 39, 43 n.10, 44, 48, 49, 50, 51, 55, 56, 58, 67, 89, 94, 170
Layzell, F. J. 137
Lee, Sir Arthur 39
Letts, bros 22
Letts, William 36, 37, 43 n.10, 46, 85 n.88, 138
Levitt, Dorothy 26, 60, 65, 75, 84, 86, 97, 106, 114, 117, 118, 120, 152, 152 n.53, 174
Linton, Arthur V. 47
Livet, Ernest 20
Llewellyn, Owen John 53, 123, 131, 132, 135, 169
Lloyd, Major Lindsay 103

Lloyd George, David 162
Lonsdale, Lord 43
Lord Advocate, The 177

Macdonald, Sir John 151
McGoun, Chip 42
McMinnies, W. G. 144
McMullen, Mr 157
'Magneto' 107
Manners, Charles 29
Mansell, Dr Horace 128
Maple, Sir John Blundell 49 n.33
Martin, G. N. 111
Martin, Myra Caroline 55
Matless, David 123
Matson, Major C. G. 18, 18 n.30, 97, 102, 103, 107, 131, 132, 146, 148
Mayhew, Mark 31, 36, 130
'Mechaniste' 118
Millar, Harold 150
Mills, G. P. 26
Minnis, John 22, 163
Mom, Gijs 8, 97, 109, 129, 134, 169
Montagu, Lord, of Beaulieu 35, 147, 148, 152 n.53, 153, 162
Moore, George 99, 100 n.45
Moore, Mrs 115
Morrison, Kathryn A. 22
Morton, H. V. 136
Munn, W. 46, 155 n.68

Napier, Montague 19, 51, 83, 91, 129
Neville, Alice Hilda 60, 118
Nicholson, T. R. 9 n.41, 15, 21, 22
Nixon, Alfred 27, 28, 33 n.126, 97, 127
Nixon, St John 17, *51*, 54, 57, 129
Nordau, Max 16
Northcliffe, Lord, *see* Harmsworth, Alfred

O'Connell, Sean 9, 108
Oddy, Nicholas 9, 115
O'Halloran, bros 154
O'Neil, Sheila 59
'Owen John', *see* Llewellyn, Owen John
Oxborrow, E. 27

Parkyns, Sir Thomas 15, 16
Patterson, Frank 124

Paulhan, Louis   146
Payne, P. L.   17
Pemberton, Max   131, 139
Pennell, Joseph   32
Pennington, Edward J.   3, 17, 55, 111
Perry, Percival   39, 43 n.10
'Pilgrim'   37, 39, 114, 115
Plowden, William   25
Porter, D.   42
Portland, Duke of   76
Prance, Henry Waymouth   117
Prince of Wales   56, 173
Prioleau, John   177 n.23

'Quartermain, Allan'   125 n.24
Queensberry, Marquis of   26

Ratliffe, J. R.   109, 120
Raven-Hill, Leonard   43
Rawlinson, Toby   44, 145
Reid, Carlton   162
Richardson, Percy   43, 44
Ritchie, Andrew   59
Rolls, Charles   35, 68, 145
'Runabout'   157
Rutter, H. Thornton   177 n.23

Salisbury, Lord   56
Salomons, David   89, 148, 151, 165, 166, 168
Sandow, Eugen   53, 54
Santos-Dumont, Alberto   144, 145, 167
Saul, S. B.   8
Savory, L.   107
Schulte, Mauritz   27
Scott-Montagu, Hon. John, *see* Montagu, Lord, of Beaulieu
Selby, James   23, 24, 47, 48
Sharp, Eleanor   46, 58, 59, 115, 116, 117, 117 n.140, 126
Sharp, George F.   144
Sharples, Rosemary   35 n.131, 58 n.92
Shorland, Frank   33, 38, 47, 48
Shrapnell-Smith, Edward   35, 144, 159
Simms, Frederick   20, 36, 37, 44, 112 n.116, 151, 151 n.40

Singer, Paris   35
Sleigh, Sir William L.   43 n.10
Smethurst, Paul   9
'Smith', *see* Cornwallis-West, Major George
Smith, G. H.   10, 22, 28, 33, 48, 115, 116, 125, 126, 127, 128, 129, 130, 140, 177
Smith, Robert J.   75, 75 n.39
Spencer, Percival   145
Spindler, G.   81
Stanley, Arthur   78
Steinbach, Susie   37
Stocks, J. W.   27, 31, 36, 43 n.10, 46, 97, 155 n.68
Stroud, Lewis   34
Sturmey, Henry   1, 2, 7, 44, 46, 89, 90, 110, 121, 149, 157, 162, 169
'Swiftsure'   38
Swindley, Harry   48

Thistlethwaite, J. B.   78
Thomlinson, Mrs   117
Thompson, F. M. L.   20, 165, 166, 172
Thornton Bridgewater, G.   29
Toms, Steven   18 n.26
Tosh, John   53
'The Tout'   38
Tracey, Dr   2, 109
Trentmann, Frank   80
Tunks, Ada   34
Turrell, Charles McRobie   44, 89, 96
Turvey, Ralph   20
Tweedy Smith, R.   25

Urry, John   8

van Hooydonk, Joseph   155
van Zuylen, Helene   60
Velasquez   41
Vincent, J. E.   125, 131

Wait, H.   46
Walford, Eric W.   163
Warwick, Countess of   59 n.102
Wellington, Frank   89, 118, 129
Wells, H. G.   162
Wertheimer, Asher   56

White, Revd H. R.   162 n.105
Whittall W.   177 n.23
Wiener, Martin   16
Wilkinson, Mr   134
Williamson, Alice   20, 41
Williamson, Charles   41
Williamson, W.   46
Williamson, writing partners   108, 123
Willoughby, Lord   26

Winchilsea, Earl of, *see* Finch-Hatton, Henry
Wodehouse, P. G.   3
Wood, Jonathan   9 n.41
Worthington-Williams, Michael   9 n.41
Wright, Sam   81
Wyatt, Horace   62
Wyeth, R. H.   80

Vehicles, cycle and motor companies

AC   131, 135, 152, 154 n.62, 158 n.82
A. C. L.   154
A. Coignet and J. Ducruzel   157 n.81
Adams-Hewitt   165
Ariel   *92*, 112 n.117, 174
Austral Cycle Company   58
Automobilette   157, 157 n.81

Baronet   15
Bédélia   157 n.81, 158
Benz   41, 59, 73, 81, 107 n.85, 110, 130, 164, 169
Bijou   154
Bollée (Leon Bollée)   *96*, 110
British Motor Syndicate   16, 44
Butler   100

Carpeviam   92, 92 n.14, 93, 98, 107
Century   46, 93
City and Suburban Electric Carriage Company   35
Civil Service Motor & Cycle Agency Ltd   108
Clement-Panhard   19
Clement-Talbot   48
Clyde   46
Cockshoots   22
Cottereau   164 n.119
Coventry Eagle   155, *156*
Crestmobile   108, 109, 120 n.157, 154
Crossley   138
Cubitt   154 n.62, 170

Daimler   1, 2, 3, 4, 20, 21, 43, 109, 138, 146, 151 n.40, 164
Darracq   44, 108, 145

De Boisse   101
De Dietrich   19, 36, 138
De Dion (De Dion Bouton)   19, 27, 45, 46, 50, 51, *52*, 56 n.76, 59, 65, 97, *102*, 118, 133, 154, 170
De Dion British and Colonial Syndicate   19, 97, 154, 154 n.68
Dodge   138
Dunhill's   37, 106, 135
   'Motorities'   37
Dunlop   48
Duryea   46

Eagle   151 n.44, 155
endurance   154
Excelsior   107

Facile   129
Ford   144, 153, 153 n.55, 161, 161 n.101, 164, 165, 168, 168 n.140

Gamage's   37, 38, 106, 135
Germain   45
Gladiator   19, 55, 59, 117
GN   158
Great Horseless Carriage Co   43
GWK   158 n.82

Hind   154
Humber   146, 154

International Motor Car Co   72, 164 n.119
Ivel   152

'Jamais Contente'   143

Kangaroo   129
Kendall   157, 157 n.81

Lacre   164 n.116
Lanchester   19, 148
Lawson Gyroscope   92, *94*
Liberty (side-carriage)   100
L. M.   157 n.81
Locomobile   46, 73
Lutzmann Benz   1

Maybach (carburettor)   17
Mercedes   39, 91
MMC   46
Morette   94, 94 n.21
Morris   144, 158 n.82, 161, 168
Mors   76
Motor Car Co   71, 72, 73
Motor Power Co   174

Napier   8, 19, 20, 28, 37, 40, *51*, 54, 55, 56, 59, 59 n.103, 61, 62, 65, 75, 80, 82, 83, 101, 116, 118, 143, 146, 147, 164 n.119, 170, 174, 176
'Napier Minor' (motor boat)   147
New Orleans   154
Nymph   108

Oldsmobile   19, 36, 138, 164 n.117
'Old Times' (coach-and-four)   23, 24, 47
Orient Express   73

Panhard and Levassor   3, 28, 44 n.14, 50, 82
  'Number 5'   89, 118
  'Number 8'   51, 82, *83*, 90
Parkyns   100
Pegasus   116
Pennington (and Baines)   3, 81, 110, *111*
Peugeot   17, 46, 50 n.40, 89, 109 n.104
Phoenix   155, 157

Portland   157, 157 n.81, 158 n.82, 164 n.119
Premier   158 n.82
Prescott   46
Princeps   46
Progress   59

Quadrant   107

Renault   148
Rex   46
Road Carrying Company   35
Rolls-Royce   44
Rothschild   44 n.14
Rover   157
Rudge   27, 50

Sabella   158 n.82
S. F. Edge (1907) Ltd   19
S. F. Edge Ltd   56, 57, 147
'Shropshire' (balloon)   145
Siddeley   86, 105, 121
Simms Motor Scout   152
Sizaire   138
Standard   135 n.95
Star   46, 73
Stellite   168, 168 n.140
Sunbeam Mabley   91, 93, 107
Swift   46, 108

Triumph   27, 71

United Motor Industries   48

Vauxhall   65, 75, 154
Velox   154
Vickers, Sons & Maxim Ltd   146, 152
V. S.   151

Wauchope's   106, 108
Werner   107
Wilbee   108
Wolseley   54, 76, 77, 118, 146

www.ingramcontent.com/pod-product-compliance
Lightning Source LLC
Chambersburg PA
CBHW072234290426
44111CB00012B/2097